W9-BWH-357

NAVEN

NAVEN

A Survey of the Problems suggested
by a Composite Picture of the Culture of a
New Guinea Tribe drawn from
Three Points of View

BY

GREGORY BATESON

SECOND EDITION

STANFORD UNIVERSITY PRESS

STANFORD, CALIFORNIA

Stanford University Press
Stanford, California
© 1958 by the Board of Trustees of the
Leland Stanford Junior University
Printed in the United States of America
Cloth ISBN 0-8047-0519-4
Paper ISBN 0-8047-0520-8
Original edition 1936
Second edition 1958
Last figure below indicates year of this printing:
81   80   79   78   77   76

To

DR A. C. HADDON, F.R.S.

PREFACE TO THE SECOND EDITION

I N this edition, the body of the book has been left unchanged. In it, a line of theoretical thought is pursued
as far as was possible in 1936. But problems similar to
those which I posed and left unsolved have since been partly
solved by cybernetics and communications theory. I have
therefore added to the book a second epilogue, in which the
theoretical positions of the book are discussed in the light of
modern developments.

What has happened has been the growth of a new way of
thinking about organization and disorganization. Today, data
from a New Guinea tribe and the superficially very different
data of psychiatry can be approached in terms of a single epistemology—a single body of questions.

We now have the beginnings of a general theory of process
and change, adaptation and pathology; and, in terms of the
general theory, we have to reexamine all that we thought we
knew about organisms, societies, families, personal relationships, ecological systems, servo-mechanisms, and the like.

GREGORY BATESON

Palo Alto, California

January 1958

FOREWORD

THE anthropologist on his travels incurs many debts. He is indebted to those who have helped him financially, he is indebted to the intellects of other anthropologists, and he is indebted to a great number of people who help him both materially and by their friendship in the countries to which he goes. For my first expedition[1] I have already expressed my indebtedness to those who helped me, and for my second expedition the list is an even longer one.

For the financing of the expedition I am indebted to the Percy Sladen Memorial Fund, to the Royal Society Government Grants Committee, and to St John's College, Cambridge, of which body I was a research Fellow, an extraordinary degree of freedom being permitted me by the Council.

Intellectually, I have to thank first and foremost my teachers, Dr A. C. Haddon, Professor A. R. Radcliffe-Brown, and Professor Malinowski. Dr Haddon first made me an anthropologist, telling me in a railway train between Cambridge and King's Lynn that he would train me and send me to New Guinea. Later his training was supplemented by Professors Radcliffe-Brown and Malinowski, and to all of these I owe much. In this Foreword, I wish to stress my admiration for Professor Malinowski's work. In the body of the book I have from time to time been critical of his views and theoretical approach. But, of course, I recognise the importance of his contribution to anthropology, and though I may think it time for us to modify our theoretical approach, the new theoretical categories which I advocate are largely built upon ideas implicit in his work.

For their criticism and for conversations which I have had with them, I wish to thank a number of people who have read parts of the manuscript of this book—Professor Radcliffe-Brown, Professor F. C. Bartlett, Dr J. T. MacCurdy, Dr Ruth Benedict, Dr E. J. Lindgren, Mr John Layard and Mr C. H. Waddington.

---

[1] *Oceania*, 1932, vol. II, nos. 3 and 4.

Of those who helped me in New Guinea, I am most indebted to Dr Margaret Mead and Dr R. F. Fortune, whom I met on the Sepik River. At a time when I was hopelessly sick of field work, I had the good fortune to meet these workers who set me a new and higher standard of work in the field. In conversations with them my approach to anthropological problems had its origin.

While we were all three of us working on the Sepik River, there arrived from America a part of the manuscript of Dr Benedict's *Patterns of Culture*, and this event influenced my thinking very profoundly. For practical help I have to thank first and foremost the Government of the mandated territory of New Guinea, especially the Government anthropologist, Mr E. W. P. Chinnery, Mr E. D. Robinson, the District Officer of the Sepik district, and Patrol Officers Keogh, Beckett, Thomas and Bloxham. Opposition of Government officials to intrusive anthropologists would be understandable, but I have not experienced it. In every contact which I had with these officials they were my friends, and often they were my hosts.

As a result of my journey to New Guinea I number many friends among its population: Mr and Mrs Mackenzie of the *Lady Betty*, Mr and Mrs E. J. Wauchope, Messrs William and Roy McGregor, Bob Overall, Bill Mason, Gibson and Eichorn. Last but not least I have to thank my native informants who were my friends, although I appeared to them ridiculous: Mali-kindjin, Tshamelwan, Tshava, Mbana, Kainggenwan, Kambwankelabi, and my cook-boy, Momwaishi.

GREGORY BATESON

St John's College, Cambridge
December 1935

TABLE OF CONTENTS

This Table is provided to help those who may be interested in dis-
secting the arguments contained in this book. Every heading in the
book is listed in the Table, and, in addition, a number of finer subdivisions, not specially marked in the text, are here given the dignity of
headings.

Chap. VI. Structural Analysis etc. (*cont.*)

Chap. VII.   The Sociology of *Naven* (*cont.*)

Fission of Iatmul communities with peripheral orientation leads to formation of new communities with the same cultural norms as the parent; fission of European systems with centripetal organisation leads to formation of daughter groups with divergent norms.

Structural and Sociological analysis has answered a number of questions about *naven*; other questions remain; exaggeration of *wau*'s behaviour; problem of the size of villages; problems of motivation; *wau*'s hypothetical desire for allegiance; answers based upon hypothetical "human nature"; difficulties in attributing affective motive.

The historian's approach to culture; *Zeitgeist* and cultural change; Configuration and the adoption of foreign traits of culture; cultural emphases, due to standardisation of individuals; standardisation either by selection or training.

Criticism of answers which invoke universal human nature; existence of opposite tendencies in human nature; need for a criterion which shall justify us in invoking one tendency rather than another; concept of standardisation provides this criterion; we must verify that the sentiments invoked are actually fostered in the culture; circular argument; its justification; *definition of ethos* (p. 118); ethos and typology; possibility that future comparative work will provide verification of ethological hypotheses.

The ceremonial house compared with a church; behaviour in the ceremonial house; self-consciousness; debates; pride in totemic ancestors; stealing of names and ancestors; ritual staged for women.

Contents

Irresponsible bullying; scarification; hazing of novices; competition
between the moieties; novices as "wives" of the initiators; ethological processes in initiation; "cutting off their own noses to spite
the other fellow's face"; a woman sees a whistle in Mindimbit, and
the secrets are therefore shown to the small boys; small boy killed
in Palimbai for insulting *wagan*, and the *wagan* are therefore shown
to women.

A captive speared; feuds; personal pride and village prosperity;
inability to revenge causes *ngglambi*; enemy corpse ritually killed;
heads and phallic standing stones; the vanquished give the names
of the dead.

The dwelling house; fishing; markets; assertive women; taking the
initiative in love; women in headhunting; seeking vengeance;
women's courage commemorated; woman's authority in the house;
double emphasis in women's ethos; the same double emphasis in
ceremonial; jolly dances for women only; innocent obscenities;
women proud in public procession; mild transvesticism.

A death at night; the women weep; a man is embarrassed; the
burial; the death of a great fighter; the men debate; they set up a
figure of the dead with symbols of his achievements; death provides
a contest for competitive boasting; later mortuary ceremonies;
mintshanggu; "quiet singing"; women's dirges stimulate the men
to caricature; pride in the presence of certain death.

Ethological contrast and Kretschmer's typology; the violent man
and the man of discretion; a "cranky" informant; types contrasted
in mythology; long noses; typology and phallic symbolism; Malikindjin, a character sketch; ambivalent feelings about him; skinny
sorcerers; native personality and culture contact; Tshimbat, a
maladjusted individual; his pig killed.

Contents

ILLUSTRATIONS

PLATES

TEXT FIGURES

NAVEN

CHAPTER I

Methods of Presentation

IF it were possible adequately to present the whole of a culture, stressing every aspect exactly as it is stressed in the culture itself, no single detail would appear bizarre or strange or arbitrary to the reader, but rather the details would all appear natural and reasonable as they do to the natives who have lived all their lives within the culture. Such an exposition may be attempted by either of two methods, by either scientific or artistic techniques. On the artistic side we have the works of a small handful of men who have been not only great travellers and observers but also sensitive writers—such men as Charles Doughty; and we have also splendid representations of our own culture in such novels as those of Jane Austen or John Galsworthy. On the scientific side we have detailed monumental monographs on a few peoples; and recently the works of Radcliffe-Brown, Malinowski and the Functional School.

These students have set themselves the same great task, that of describing culture as a whole in such a manner that each detail shall appear as the natural consequence of the remainder of the culture. But their method differs from that of the great artists in one fundamental point. The artist is content to describe culture in such a manner that many of its premises and the inter-relations of its parts are implicit in his composition. He can leave a great many of the most fundamental aspects of culture to be picked up, not from his actual words, but from his emphasis. He can choose words whose very sound is more significant than their dictionary meaning and he can so group and stress them that the reader almost unconsciously receives information which is not explicit in the sentences and which the artist would find it hard—almost impossible—to express in analytic terms. This impressionistic technique is utterly foreign to the methods of science, and the Functional School have set out to describe in analytic, cognitive terms the whole interlocking—almost living—nexus which is a culture.

Very properly and naturally they have paid greatest attention to those aspects of culture which lend themselves most readily to description in analytic terms. They have described the structure of several societies and shown the main outlines of the pragmatic functioning of this structure. But they have scarcely attempted the delineation of those aspects of culture which the artist is able to express by impressionistic methods. If we read *Arabia Deserta*, we are struck by the astonishing way in which every incident is informed with the emotional tone of Arab life. More than this, many of the incidents would be impossible with a different emotional background. Evidently then the emotional background is causally active within a culture, and no functional study can ever be reasonably complete unless it links up the structure and pragmatic working of the culture with its emotional tone or ethos.[1]

The present work is a description of certain ceremonial behaviour of the Iatmul people of New Guinea in which men dress as women and women dress as men, and an attempt—crude and imperfect, since the technique is new—to relate this behaviour, not only to the structure and pragmatic functioning of Iatmul culture, but also to its ethos.

This enquiry will involve me in a perhaps tedious discussion of abstractions, and to those who have some difficulty with, or a distaste for, epistemology I would recommend that they read first the descriptive chapters, especially those on the ethos of Iatmul culture, so that a preliminary study of concrete examples may make clear my abstractions. Others may find that the Epilogue, in which I have recounted some of the

---

[1] At a later stage I shall examine and attempt to define these concepts of Structure, Function and Ethos in a more critical manner, but since the concept of Ethos is still unfamiliar to many anthropologists, it may be well to insert here the definition which the *Oxford English Dictionary* gives of the word. The definition reads: " 1. The characteristic spirit, prevalent tone of sentiment of a people or community; the 'genius' of an institution or system." The earliest example of the use of *ethos* in this sense is quoted from Palgrave, 1851. I must ask classical scholars to forgive the plural "ethoses" which I have coined in preference to "ethe". The latter could never be English.

theoretical errors into which I fell in the course of my enquiry, will help them to understand my present theoretical position.

At this early stage, I wish to make it perfectly clear that I do not regard Ritual, Structure, Pragmatic Functioning and Ethos as independent entities but as fundamentally inseparable aspects of culture. Since, however, it is impossible to present the whole of a culture simultaneously in a single flash, I must begin at some arbitrarily chosen point in the analysis; and since words must necessarily be arranged in lines, I must present the culture, which like all other cultures is really an elaborate reticulum of interlocking cause and effect, not with a network of words but with words in linear series. The order in which such a description is arranged is necessarily arbitrary and artificial, and I shall therefore choose that arrangement which will bring my methods of approach into the sharpest relief. I shall first present the ceremonial behaviour, torn from its context so that it appears bizarre and nonsensical; and I shall then describe the various aspects of its cultural setting and indicate how the ceremonial can be related to the various aspects of the culture.

Throughout this analysis I shall confine myself to synchronic explanations of the phenomena, that is to say, to explanations which invoke only such other phenomena as are now present in Iatmul culture, or which, like head-hunting, have disappeared so recently that they can be considered as part of the natural setting of the ceremonies as I collected them. I shall not enquire what either the ceremonies or their cultural setting may have been like in the past. In my use of causal terminology I shall be referring to *conditional* rather than to *precipitating* causes. Thus in a synchronic study of a fire I should say that the fire burns because there is oxygen in the room,[1] etc., but I should not enquire how the fire was first ignited.

---

[1] I have endeavoured to avoid phrasings reminiscent of the statement, "the square of the velocity is a cause of the kinetic energy". But when I was first working out the lines of thought upon which this book is based I not infrequently fell into this verbal error and it is likely that in some cases the mistake has passed uncorrected.

It is of course true that the ceremonies have had a history and it would no doubt be possible to speculate about that history. But such is not my purpose. I shall be content only to show some of the sorts of functional relationship which exist between the ceremonies and the remainder of the contemporary culture of the Iatmul. Perhaps, in the future, a clearer understanding of the synchronic aspects of society will enable us to isolate and define the diachronic, the processes of cultural change.

The Iatmul people[1] live on the middle reaches of the Sepik River in the Mandated Territory of New Guinea. They are a fine, proud, head-hunting people who live in big villages with a population of between two hundred and a thousand individuals in each village. Their social organisation, kinship and religious systems are developed to an extreme of complexity. The community is subdivided into groups according to two independent systems with very little congruence between one system and the other. On the one hand there is a division into two totemic moieties which are further subdivided into phratries and clans; and on the other hand there is a division into two cross-cutting pairs of initiatory moieties which are subdivided into age grades. None of these groups are strictly exogamous. Membership of all the groups is determined by patrilineal descent.

In spite of this strong emphasis upon patriliny, the people pay a great deal of attention to kinship links through the mother or sister, and both the patrilineal and matrilineal links are preserved in a classificatory system through many generations. Thus the term *wau* (mother's brother) and *laua* (sister's child, m.s.) are used not only between own mother's brother and sister's children, but also in a classificatory way, so that the term *wau* includes such relatives as mother's mother's sister's son (cf. Fig. 3, p. 94) even though all three of the

[1] A preliminary account of this people, based on the results of my first expedition, has been published in *Oceania* (1932, vol. II, nos. 3 and 4). The present paper is based upon the work of this first expedition as well as on my second expedition when I spent a further fifteen months among the Iatmul. My first expedition was financed by Cambridge University with a grant from the Anthony Wilkin Fund. The second was financed jointly by the Royal Society and the Percy Sladen Trust.

intervening women through whom the kinship is traced have married into different clans. It is with the relations between classificatory *waus*[1] and *laua*s that this book chiefly deals.

[1] In the Iatmul language there is no suffix attached to substantives to indicate plurality. I have therefore used the English suffix, -s, for this purpose. In place of the Iatmul possessive suffix, -*na*, I have used the ordinary English -'s. Native words are in all cases italicised but the English suffixes remain in ordinary type.

CHAPTER II

The Naven *Ceremonies*

OCCASIONS ON WHICH *NAVEN* ARE PERFORMED

T HE ceremonies are called *naven* and are performed in celebration of the acts and achievements of the *laua* (sister's child). Whenever a *laua*—boy or girl, man or woman—performs some standard cultural act, and especially when the child performs this act for the first time in its life, the occasion may be celebrated by its *wau*. The possible occasions for the performance of *naven* are very numerous and very frequent. In the case of a boy the list of acts and achievements which may be celebrated in this way is a long one, and may conveniently be divided into five categories:

1. Major achievements which, though greeted with a more elaborate *naven* upon their first performance, are also received with some show of *naven* behaviour every time they occur. Of these the most important is homicide. The first time a boy kills an enemy or a foreigner or some bought victim is made the occasion for the most complete *naven*, involving the greatest number of relatives and the greatest variety of ritual incidents. Later in his life when the achievement is repeated, there will still be some *naven* performance on the part of the *wau*, but the majority of the ritual incidents will probably be omitted. Next to actual homicide, the most honoured acts are those which help others to successful killing. The man standing on the bow of a war canoe does not carry a spear-thrower but carries instead a very light paddle set in a long bamboo shaft; with this he wards off darts from the spear-throwers of the enemy. This man may be honoured with *naven* for any kills effected by members of his canoe. Another act contributory to killing which may be honoured is the enticing of foreigners into the village so that others may kill them. Far behind these in importance come such achievements as the killing of a large crocodile, the killing of a wild

pig, the spearing of a giant eel, etc.—achievements which are still sufficiently important whenever they are repeated to stimulate the *wau* at least to a ritual greeting and perhaps to the throwing of a cloud of lime at the *laua*.

2. Minor cultural acts which are celebrated only upon the first occasion of their achievement. Actually it would perhaps be correct to say that the first performance of *any* cultural act may be the excuse for a *naven*. An informant in Mindimbit village, however, gave me a long list, and I have added to it one or two other acts which I know may be celebrated with *naven*. This list, though of course it is not complete, is yet worth reproducing, because it illustrates the sort of act which may be noticed by the *wau*. It includes: killing any of the following animals—birds, fish, eel, tortoise, flying fox; planting any of the following plants—yams, tobacco, taro, coconut, areca, betel, sago, sugar-cane; spotting an opossum in the bush; felling a sago palm, opening it and beating sago; using a spear-thrower; using a throwing stick to kill a bird; using a stone axe (or nowadays a steel knife or axe); sharpening a fish spear; cutting a paddle; making a canoe; making a digging stick; making a spear-thrower; incising patterns on a lime gourd; plaiting an arm band; making a shell girdle; beating a hand drum; beating a slit-gong; blowing a trumpet, playing a flute, beating the secret slit-gongs called *wagan*; travelling to another village and returning; acquiring shell valuables; buying an axe, knife, mirror, etc.; buying areca nut; killing a pig and standing a feast.

3. Acts characteristic of *laua*. The relationship between *laua* and *wau* is marked by various forms of characteristic behaviour—acts which each performs in reference to the other. Such acts may be described as duties, services, or privileges, and it is not at all easy in any given case to decide which of these terms should be applied. In general, whenever the *laua* performs any conspicuous act which is characteristic of his position as *laua*, the *wau* will respond with some show of *naven* behaviour. These acts include such ceremonials as the exhibition of totemic ancestors of *wau*'s clan, dancing in masks which represent these ancestors and carving their

images on the posts which are to stand in the section of the ceremonial house which belongs to the *wau*'s clan. Several of the musical activities mentioned above—e.g. beating secret slit-gongs and playing flutes, may also fall within this category of services to the *wau*'s clan. A man may blow the flutes of his own clan and his musical accomplishment be greeted by *naven* upon the first occasion, but it is also his duty or privilege to blow the totemic flutes of his *wau*'s clan on certain special occasions; and this act, like other honouring of the *wau*'s totems, will be greeted with *naven* behaviour upon every occasion when it is repeated. Besides ritual duties and privileges, the *laua* will help his *wau* in such labours as house building; and in the formal debating in the ceremonial house, the *laua* will often speak on the side of his *wau*'s clan. All such acts will be hailed by the *wau* whenever they occur with gestures and exclamations reminiscent of *naven*.

4. Boasting in the presence of the *wau*. It is correct for a boy to boast in the presence of his *wau*, but the latter may resent this behaviour if it is carried to excess; and he will, in this event, make a gesture suggestive of turning his buttocks to his *laua*. I have never seen the complete gesture of rubbing the buttocks on the *laua*'s shin, the climax of the *naven*, carried out in reply to *laua*'s boasts; usually the threat is sufficient to curb the young man's tongue. But it is generally stated that, when exasperated, a *wau* may complete the gesture and, by so doing, involve the *laua* in a presentation of valuables to the *wau*. It is probable that this boasting in the presence of the *wau* should be classified with the other acts characteristic of the *laua*. But the case differs from the others in that the *naven* behaviour of the *wau* is carried out in anger or annoyance. In the other cases the *naven* behaviour is on the whole a method of complimenting or congratulating the *laua* upon his achievements.

5. Changes in social status. Such events in the boy's life as: the boring of his ears, the boring of his nasal septum, his initiation, his marriage, his becoming possessed by a shamanic spirit, may all be celebrated with *naven* when they occur. I feel some hesitation in applying the term *change of status* to these events, since the culture does not clearly emphasise

the concepts of status and promotion. Even in the graded
initiatory system, the event which the *wau* celebrates is not
the promotion of the boy from one grade to the next, but the
completion of the ceremony of initiation which may take
place years before the general shifting of age-grade membership. In any case it must be clearly understood that the
naven ceremonies are in no sense *rites de passage*, although
they may celebrate or emphasise the fact that such rites have
been performed. The actual rites of initiation are performed
by the elder age grade within a system of social groupings
and moieties which is to a great extent independent of clans
and totemism. The fact that the *rite de passage* has been
performed is afterwards celebrated by the *wau* within the
different social grouping which is organised on the basis of
the family and totemic clans.

The other most important events in a man's life—his birth
and his death—are neither of them celebrated by *naven*. In
the case of birth, the *wau* will go to the child soon after it is
born and present it with a coconut and a personal name which
refers to the totemic ancestors of the *wau*'s clan. Though
I have never seen this done, I believe that the *wau* might
well exclaim: "*Lan men to!*" (husband thou indeed!) when
the baby gripped his fingers. This exclamation, as we shall
see later, is one of the characteristic details of *naven* behaviour.
We shall see too that the giving of the special name is an act
which demonstrates the existence of the *wau-laua* relationship; and that the coconut is the first in a long series of gift
exchanges which will accompany *naven* and other ceremonies.
In the case of death, there is again no *naven*, but the classificatory *wau* plays an important part in the mortuary ceremonies, and finally claims the dead man as in some special
sense a member of the maternal clan, pulling the figure which
represents him towards himself with a hook. In the land of
the dead the ghost will henceforth live under the names which
have been given to him by his *wau*.

The event of marriage may be celebrated not only by *wau*,
but also, I believe, by *tawontu* (wife's brother). In one of
the myths which I collected, there occurs a casual mention
of the fact that a bride's own brother rubs his buttocks on

the bridegroom's shin. I do not know of any other occasion which is celebrated in this way by *tawontu*.

There are more *naven* and more occasions for *naven* in the life of a boy or a man than in the life of a girl, but the achievements of a girl may also be occasions for *naven*. The list for a girl includes: catching fish with hook and line; collecting mayflies; washing sago; cooking sago pancakes; cooking sago paste; making a fish trap, a rain cape or a mosquito bag; and bearing a child. These events all fall into the category of cultural acts which are celebrated at their first accomplishment. Besides these there are two other events, initiation and dancing in the *tshugukepma* dances, both of which may be celebrated with *naven*. The initiation of girls is distinct from female puberty ceremonies and consists of scarification and the showing of flutes. It is a simplified version of the initiation of boys and is carried out by the men of the elder age grade. The ceremony is performed on only a very few women. The *tshugukepma* ceremonies are occasions on which the *laua*s exhibit the totemic ancestors of the *wau*'s clan and dance in masks which represent these ancestors, so that they fall into the category of "acts characteristic of *laua*". I do not know anything about girls boasting in the presence of their *wau*s; in any case *wau*'s gesture of rubbing his buttocks on *laua*'s shin is specifically a part of *naven* for a boy, and an analogue of this gesture in the ceremonies for a girl may be seen in the pantomime of her birth from the belly of *wau*.

MATERIALS UPON WHICH THE DESCRIPTION IS BASED

Although the list of possible occasions is so long, in actual practice large *naven* ceremonies are not very often performed, and their frequency is perhaps limited by the expense which they involve. In most instances, if the *laua*'s achievement is brought to the notice of the *wau*, the latter only exclaims: "*Lan men to!*" (husband thou indeed!), throws some lime over his *laua* and hails him ceremonially with a string of names of ancestors of *wau*'s clan. Besides these instances in

which the *naven* ritual is reduced to mere gestures and exclamations, there are many *naven* which are only celebrated
on a small scale. It is hard for me to judge of the frequency
of these small *naven* because I may often have heard nothing
of such occurrences even though they took place while I was
in the village. I have only witnessed five *naven* in which any
of the ritual was carried out. Of these, one was in the village
of Mindimbit, for a group of children who had been out
working sago in the swamps and had returned, the little boys
having accomplished the felling of the palm and the beating
of the pith and the little girls having washed the resulting
pulp to get the sago (cf. the myth quoted on p. 36). I witnessed
two *naven* in Kankanamun village, one of which was for a
boy who had made a canoe and the other for a young man
who had killed a pig and stood a feast. In both of these
ceremonies only the women (mothers, fathers' sisters, and
elder brothers' wives) took part, the *waus* apparently ignoring
the occasions. I witnessed one *naven* in Palimbai for a big
new canoe; in this only two *waus* took part and the occasion
was ignored by the women. Lastly, I witnessed an incident
in Malinggai in which a *wau* dashed into the midst of a
tshugukepma dance and rubbed his buttocks on the shin of
a male *laua* who was impersonating one of the *wau*'s clan
ancestors. Besides these small *navens*—none of which was
at all elaborate—I obtained good descriptions of a larger
naven which took place in Mindimbit for a little girl who
had caught a fish with hook and line, and I collected, also in
Mindimbit, good descriptions of the elaborate and varied
naven ceremonies which used to be held for the successful
homicide. My account of *naven* is based on this material
together with scattered references to *naven* in the conversations of informants and in mythology.

The four villages, Mindimbit, Kankanamun, Palimbai and
Malinggai, are closely related in language and culture, and
references to the small local differences in social organisation
and behaviour will be inserted when relevant.

DESCRIPTION OF THE CEREMONIES

The outstanding feature of the ceremonies is the dressing of men in women's clothes and of women in the clothes of men. The classificatory *wau* dresses himself in the most filthy of widow's weeds, and when so arrayed he is referred to as "*nyame*" ("mother"). Plates II to IV show two classificatory *wau*s arrayed for the *naven* of the young man in Palimbai who had made a large canoe for the first time in his life. They put on the most filthy old tousled skirts such as only the ugliest and most decrepit widows might wear, and like widows they were smeared with ashes. Considerable ingenuity went into this costuming, and all of it was directed towards creating an effect of utter decrepitude. On their heads they wore tattered old capes which were beginning to unravel and to fall to pieces with age and decay. Their bellies were bound with string like those of pregnant women. In their noses they wore, suspended in place of the little triangles of mother-of-pearl shell which women wear on festive occasions, large triangular lumps of old sago pancakes, the stale orts of a long past meal.

In this disgusting costume and with absolutely grave faces (their gravity was noted with special approbation by the bystanders), the two "mothers" hobbled about the village each using as a walking stick a short shafted paddle such as women use. Indeed, even with this support, they could hardly walk, so decrepit were they. The children of the village greeted these figures with screams of laughter and thronged around the two "mothers", following wherever they went and bursting into new shrieks whenever the "mothers", in their feebleness, stumbled and fell and, falling, demonstrated their femaleness by assuming on the ground grotesque attitudes with their legs widespread.

The "mothers" wandered about the village in this way looking for their "child" (the *laua*) and from time to time in high-pitched, cracked voices they enquired of the bystanders to learn where the young man had gone. "We have a fowl to give to the young man" (cf. Plate II). Actually the *laua* during this performance had either left the village or hidden himself. As soon as he found out that his *wau*s were going to shame themselves in this way, he went away to avoid seeing the spectacle of their degraded behaviour.

If the *wau* can find the boy he will further demean himself by rubbing the cleft of his buttocks down the length of his *laua*'s leg, a sort of sexual salute which is said to have the effect of causing the *laua* to make haste to get valuables which he may present to his *wau* to "make him all right".[1] The *laua* should, nominally at least, fetch valuables according to the number of times that the *wau* repeats the gesture—one shell for each rubbing of the buttocks.

The *wau*'s gesture is called *mogul nggelak-ka*. In this phrase the word *mogul* means "anus", while *nggelak-ka* is a transitive verb which means "grooving", e.g. *ian nggelak-ka* means to dig a ditch. The suffix *-ka* is closely analogous to the English suffix, -ing, used to form present participles and verbal nouns.

This gesture of the *wau* I have only seen once. This was when a *wau* dashed into the midst of a dance and performed the gesture upon his *laua* who was celebrating the *wau*'s ancestors. The *wau* ran into the crowd, turned his back on the *laua* and rapidly lowered himself—almost fell—into a squatting position in such a way that as his legs bent under him his buttocks rubbed down the length of the *laua*'s leg.

But in the particular *naven* which I am describing, the two *wau*s did not find their *laua* and had to content themselves with wandering around the village in search of him. Finally they came to the big canoe which he had made—the achievement which they were celebrating. They then collapsed into the canoe and for a few moments lay in it apparently helpless and exhausted (Plate IV), with their legs wide apart in the attitudes which the children found so amusing. Gradually they recovered and picked up their paddles, and sitting in the canoe in the bow and stern (women sit to paddle a canoe, but men stand), they slowly took it for a short voyage on the

---

[1] This is the pidgin English translation of an Iatmul phrase, *kunak-ket*. The suffix *-ket* is purposive; and the word *kunak* means to "make ready", "repair", or "propitiate".

lake. When they returned they came ashore and hobbled off.
The performance was over and they went away and washed
themselves and put on their ordinary garments. The fowl
was finally given to the *laua* and it became his duty to make
a return present of shell valuables to his *wau* at some later
date. Return presents of this kind are ceremonially given,
generally on occasions when some other dances are being
performed. The shells are tied to a spear and so presented
to the *wau*.

In more elaborate *naven*, especially those in which women
play a part, there is a classificatory spreading of ritual behaviour not only to cause the classificatory relatives of the
actual *laua* to perform *naven* for him, but also to cause persons not otherwise involved to adopt *naven* behaviour towards
other individuals who may be identified in some way with
the actual *laua*. For example, the characteristic *naven* behaviour of the elder brother's wives is the beating of their
husband's younger brother when his achievements are being
celebrated. Owing to the classificatory spreading of the *naven*
not only does the boy who has worked sago get beaten by
his elder brother's wives, but also the boy's father's elder
brothers' wives get up and beat the father. Further, men
other than the performing *wau*s may take the opportunity
to make presentations of food to their various *laua*s.

Some of this classificatory spreading occurred in the *naven*
in Mindimbit which was celebrated for the children who
had worked sago for the first time. In this ceremony only
female relatives performed and their costume was in sharp
contrast to that of the *wau*s described above. The *wau* dresses
himself in the filthiest of female garments, but the majority
of the women, when they put on the garments of men, wear
the smartest of male attire. The female relatives who performed were sisters (*nyanggai*), fathers' sisters (*iau*), elder
brothers' wives (*tshaishi*), mothers (*nyame*), and mothers'
brothers' wives (*mbora*)—all these terms being used in a
classificatory as well as in the narrow sense.

Of these relatives the sisters, fathers' sisters and elder
brothers' wives dressed as men, borrowing the very best of
feather headdresses and homicidal ornaments from their

menfolk (husbands or brothers or fathers). Their faces were painted white with sulphur, as is the privilege of homicides, and in their hands they carried the decorated lime boxes used by men and serrated lime sticks with pendant tassels whose number is a tally of men killed by the owner. This costume was very becoming to the women and was admired by the men. In it the women were very proud of themselves. They walked about flaunting their feathers and grating their lime sticks in the boxes, producing the loud sound which men use to express anger, pride and assertiveness. Indeed so great was their pleasure in this particular detail of male behaviour that the husband of one of them, when I met him on the day following the performance, complained sorrowfully that his wife had worn away all the serrations on his lime stick so that it would no longer make a sound.

The mothers and mothers' brothers' wives wore different types of costume. The mother stripped off her skirt but did not put on any male ornaments, and the mothers' brothers' wives put on filthy widow's weeds, like those of the *wau*, described above.

With these modifications of costume there goes also a modification of the kinship terms used for these women. Thus:

iau (father's sister) becomes "*iau-ndo*" or "*nyai*'" (literally "father").

tshaishi (elder brother's wife) becomes *tshaishi-ndo* or *nyamun* (literally "elder brother").

mbora (mother's brother's wife) becomes *mbora-ndo*.

nyanggai (sister) becomes *nyanggai-ndo*.

In these special terms the suffix *-ndo* means "man". The mothers, who alone of the female relatives concerned in *naven* put on no male attire, are still described as *nyame* (mother).

The children's canoe came back from the sago swamps late in the morning and as soon as it was sighted from the banks of the river, word was shouted in to the village which lies back from the river on a little lake. The women assembled on the shores of the lake and when the canoe entered, they swam out to greet and splash the children, as is done when a canoe returns from a successful head-hunting raid. When the children had landed, the village appeared to go mad for awhile; fathers' sisters and elder brothers' wives dashing about searching respectively for their various brothers' children and husbands' younger brothers in order to beat them. The men who were expecting to be beaten did their best to avoid the ceremony by skulking in the ceremonial houses, but the women on these occasions have unusual licence (perhaps because they are in men's costume). Whenever a *iau* saw her *kanggat* (brother's child, w.s.) or a *tshaishi* saw her *tshuambo* (husband's younger brother) in one of the ceremonial houses, she dashed into the usually forbidden spot, stick in hand, and gave him several good blows; and, if he ran away, she chased him, beating him as he ran. On a similar occasion in Kankanamun, when the women were celebrating *naven* for a young man who had killed a pig and was standing a feast, they hesitated to enter a ceremonial house in which a debate was in progress. The men stopped the debate for their benefit and the whole crowd of them danced into the ceremonial house in a column. Arrived there, the column broke up and each woman went and beat her appropriate relatives.[1]

On the occasion in Mindimbit which I am describing, I did not see any activity of the mothers and sisters. The activity of the other women dressed as men continued off and on throughout the rest of the day. In the evening, the women had a small dance by themselves. It is the custom of the men to take off their pubic aprons after dark and the women accordingly stripped off the flying-fox-skin aprons, which they had borrowed from their husbands and brothers, and danced with their loins uncovered, still wearing their splendid feather headdresses and ornaments. There was no apparent embarrassment of the men at this exposure in their women-

---

[1] To this account of the beatings, two points may be added which I did not observe at the time but of which I was later informed. It is said that the actual father of the hero would only be slightly beaten by his *tshaishi* but that the father's younger brother (*tshambwi-nyai'*) would be severely beaten. I was also told that on these occasions the *tshaishi* will exclaim "*tshuambo-ket wonggegio*"—(I will rape my young brother).

folk—but the older men were rather shocked at this lack of embarrassment. The dance was held close to one of the smaller ceremonial houses and one of the old men remarked to me in scandalised tones that it was "shocking" to see the younger men crowding to this ceremonial house, where they had no business but a good view of the women's dance. It was the ceremonial house of a particular clan, but that evening it was full of young men, members of every clan in the village.

In the case of the *naven* in Mindimbit for the little girl who had caught a fish with hook and line, the performance was still more elaborate, not because the achievement was very important but perhaps because the *wau*s were anxious to obtain shell valuables by presenting pigs to their *laua*s. From the accounts which I collected, it appears that in this *naven* both male and female relatives took part and eight pigs were killed. Besides the little girl who had caught a fish (a month or two before the ceremony), two other little girls were honoured, and the resulting *naven* spread to almost every individual in the village. One pig was even presented to a classificatory *laua* in the next village.

Four *wau*s were dressed as "mothers" and their skirts were tucked up so as to expose their genitals. Three of these *wau*s are described as carrying the little girls "on their heads"— presumably carrying girls in the position in which mothers habitually carry their children astride on the shoulder (cf. Plate V B). The fourth *wau* was also dressed as a mother but wore no skirt. He was tied down on some sort of "bed" or stretcher, on which he was lifted and violently swung by a number of men who, while they rocked the stretcher, sang songs of the *wau*'s clan. The little girl who had caught the fish was placed on the belly of this *wau*, her father meanwhile standing by with an adze to which he had tied a mother-of-pearl crescent. This adze he gave into the little girl's hands and with it she cut the bonds which held her *wau* to the stretcher. She then gave the decorated adze to the *wau* and with it he raised himself to a sitting position on the stretcher, supporting himself by means of the adze. Similarly the fathers of the other little girls presented valuables to the girls' *wau*s and at the same time loosed the *wau*s' skirts. The *wau*s then resumed their normal male attire.

During the above performance a *mbora*, the wife of the *wau* who was tied to the stretcher, danced with her skirts tucked up to expose her genitals. She wore a string bag over her head and face, and carried a digging stick, holding it horizontally behind her shoulders with her hands raised to hold it on each side. At the conclusion of the dance she was presented with a mother-of-pearl crescent and three *Turbo* shells by the father of the little girl.

The whole of this ritual pantomime acted by the *wau* appears to me to be a representation of the birth of the little girl from the belly of her mother's brother, though none of my male informants (in Kankanamun where I made enquiries) had ever heard of the custom of tying the woman in labour on a stretcher. The dance of the *mbora* (mother's brother's wife), with her arms stretched back behind her head, also probably represents the position of a woman in labour.

A general presentation of food and valuables followed this ceremony. Eight pigs were killed and presented. Of these one was given by her *wau* to the little girl who had caught a fish, and she ceremonially stepped upon it. Of the remaining pigs, three were given by *wau*s to classificatory *laua*s, including one pig which was given to the donor's sister's husband's father,[1] in the next village. One of the other pigs was given by a woman to her husband's sister's child (i.e. *mbora* gave to *nasa*, husband's sister's child). One was given to a classificatory sister's husband (i.e. to the potential father of a *laua*), and one was given to a widowed classificatory sister.[2]

---

[1] This relative also is called *laua*. The Iatmul kinship system is characterised by an alternation of generations similar to that found in many Australian systems. Theoretically a man is re-incarnated in his son's son; and a woman in her brother's son's daughter. In general the kinship term which is appropriate for a given relative will also be appropriate for any other relative of the same sex in an analogous position two patrilineal generations away. Thus the term *nyamun* is used both for elder brother and for father's father. The term *nyame* is used both for mother and for son's wife. In the present instance the term *laua* is used both for sister's husband's son and sister's husband's father.

[2] According to the collected account, the eighth pig was given by a man to his own mother's brother's son. This is anomalous and probably either the account is at fault or the two persons concerned were also related in an inverse way through some other route.

We have seen above that the father of the *laua* provided
the valuables which the little girl gave to her *wau*. In the
general presentation of food and valuables this pattern is
further extended so that the exchanges actually take place
between the relatives of wives on the one hand and the relatives of husbands on the other. The wife's clan and her
classificatory brothers, etc., give pigs to the husband's son or
to the husband's father or to the husband. This system is
summed up in the native language by saying that the *wau-
nyame nampa* (mother's brother and mother people) give pigs
to the *lanoa nampa* (husband people) or *laua nyanggu* (sisters'
children people)—these two terms being almost synonymous
(cf. p. 93). In the return presentation of valuables, the
classificatory brothers and other relatives of the recipients
of pigs contributed to the accumulation of valuables which
made up the return gifts, so that the list was exceedingly
elaborate and involved a great number of people, related in
the most various ways to the original donors of the pigs.

The last *naven* to which I shall refer is that which used to
be celebrated for the successful homicide when he had accomplished his first kill. The account which I have of these ceremonies repeats some of the material which I have already
given, but it also describes four other pantomime incidents
which may be celebrated on such occasions. The incidents
are apparently independent one of the other, but all of them
involve the same types of transvesticism which I have described above. The only exception to this is the costume of
mbora, who though she wears ragged female attire in the
first incident puts on (probably ragged[1]) male attire in the
second and third.

In the first incident, the *mbora* dances in bedraggled skirt
with her head inside a fishing net. She wears the enemy
head suspended from her neck, and with her raised hands she
holds a digging stick behind her shoulders. Another *mbora*
might have the jaw bone from the enemy head hung around
her neck. This dance continues until sunset when the *nasa*

---

[1] Unfortunately I do not know whether the *mbora* wears smart or
bedraggled costume when dressed as a male—but I suspect that she, like
her husband, wears filth.

(husband's sister's son) brings a *Turbo* shell tied to a spear and, presenting this, looses the old skirt from his *mbora*.

In the second incident, the *iau-ndo* (father's sisters in male costume) carry in their hands a feather ornament such as the homicide will wear in his headdress. The *mbora-ndo* (mother's brother's wives in male costume) lie down on the ground and the *iau-ndo* go stepping across them carrying the feathers. The *mbora-ndo* then snatch the feathers and run off with them.

In the third incident, the *wau* puts on a skirt and fixes an orange-coloured fruit, *mbuandi* (*Ervatamia aurantiaca*), in his anus and goes up the ladder of a house displaying this as he climbs. At the top of the ladder he goes through the actions of copulation with his wife, who is dressed and acts as the male. The *laua* is much ashamed at this spectacle and the *laua*'s sister would weep at it. The orange fruit represents an anal clitoris, an anatomical feature frequently imagined by the Iatmul and appropriate to the *wau*'s assumption of grotesque feminity. My informant told me that, after the *mbora*, acting as male, had copulated with the *wau*, the other women would all follow suit—and we may imagine a scene of considerable confusion around the unfortunate *wau*.

In the fourth incident, a large pear-shaped prawn trap with lobster pot entrance is put on the ladder of a house. All the women of the village lie down naked on the ground, side by side in front of the ladder. The killer steps over the women to go up into the house. As he walks, he is ashamed to look at their genitals; he therefore goes with his head up, feeling his way with his feet. On this occasion the women would say of the vulva: "That so small a place out of which this big man came." The only women over whom he would not step are his sister and his wife. His sister would accompany him as he steps. She shows no modesty but attacks the vulvae with her hands as she passes—especially that of *tshaishi*, the elder brother's wife. Seeing this, she would exclaim: "A vulva!" But *tshaishi* would reply: "No! A penis!"

After this the *tshaishi* sings a comic song, beating time with a coconut shell. The text of this song, like that of all

Iatmul songs, is built upon a series of totemic names, in this
case, names of the fish trap set on the ladder.

> This fish trap, Alie-namak[1] fish trap.
> Go! you will never shoot it.
> Go! you will never spear it.
> This is Woli's[2] fish trap.
> This is Tanmbwa's[3] fish trap.
> Go! You will never shoot it.
> Go! You will never spear it. Etc.

This is repeated substituting other names for Alie-namak in
the first line. At the close of the song the killer spears the
trap and goes up into the house.

In this last incident, the fish trap is certainly a symbol of
the vulva, and this symbol occurs also in esoteric mythology,
together with the complementary symbol, the eel, which gets
caught in the trap.

Before attempting to account for the details of the ceremonial, we may conveniently summarise the *naven* behaviour
of the various relatives:

Wau (mother's brother) wears grotesque female attire;
offers his buttocks to male *laua*; in pantomime gives birth
to female *laua* who looses his bonds; supports himself on
the adze presented by her; presents food to *laua* of either
sex and receives in return shell valuables; acts as female in
grotesque copulation with *mbora*. These ceremonial acts may
be performed either by own *wau* or classificatory *wau*—most
usually the latter.

Mbora (mother's brother's wife) wears either grotesque
female attire or (probably grotesque) male attire; dances with
digging stick behind her head; takes male part in mimic
copulation with *wau*; like *wau* she presents food to the hero
of the *naven* and receives valuables in return.

Iau (father's sister) wears splendid male attire; beats the
boy for whom *naven* is celebrated; steps across his prostrate
mbora; participates in mimic contest between *mbora* and *iau*

---

[1] Alie-namak: this is a totemic name for a fish trap, an ancestor of
Mandali clan. The meaning of the word, *alie*, is unknown to me. The word,
namak, is an old term for *namwi*—fish trap.

[2,3] Woli and Tanmbwa: these are two names for the East Wind and
are used in the jargon of shamans to refer to women. Each of these lines
therefore means: "This is a woman's fish trap."

in which the former snatch the feather headdress from the latter.

Nyame (mother) removes her skirt but is not transvestite; lies prostrate with other women when the homicide steps across them all.

Nyanggai (sister) wears splendid male attire; accompanies the homicide, her brother, when he steps across the women; he is ashamed but she attacks their genitals, especially that of *tshaishi*; weeps when *wau* displays anal clitoris.

Tshaishi (elder brother's wife) wears splendid male attire; beats husband's younger brother and her vulva is attacked by his sister.

Tawontu (wife's brother, i.e. *wau* of ego's children) presents food to sister's husband (*lando*) and receives valuables in return. *Tawontu* may, I believe, rub his buttocks on *lando*'s shin, when the latter marries *tawontu*'s own sister.

CHAPTER III

The Concepts of Structure and Function

STRUCTURE

I HAVE described the *naven* ceremonies with a minimum of references to the culture in which they are set; and any attempt to *explain* them must take the form of relating the ceremonies to their setting. Even to construct an historical theory of their origin would require an exhaustive examination of the remainder of the culture, and this is still more true of any theorising which would account for the *naven* in terms of the structure or of the functioning mechanisms of the culture. Since I shall try to give analyses of the position of *naven* from both structural and functional points of view, it will be well to state definitely what I mean at this stage[1] by the two concepts, Structure and Function.

The anthropologist in the field collects details of culturally standardised behaviour. A large part of this material takes the form of native statements *about* behaviour. Such statements may be seen as themselves details of behaviour; or, more cautiously, we may regard them as true and supplementary to the anthropologist's account of behaviour which he has witnessed.

We can view any such detail from a structural point of

---

[1] This chapter contains a statement of my theoretical position when I was writing the body of this book. The various descriptive chapters were a series of experiments in the use of the different abstractions which are here discriminated and defined. On the whole these abstractions have stood the test of experiment, and their use has led to generalisations which I have thought interesting.

The meaning which I attach to the terms *affective* and *cognitive* is here left vague, and the whole book was written with a vague use of these terms. I hoped that experiments in the description of culture might clear up this matter which the study of individual psychology has left obscure. In the Epilogue from a comparison of my various methods I have gone on to attempt the solution of this problem, and the position at which I finally arrive is slightly different from that which is outlined in this chapter. To re-write my statement of working hypotheses in terms of the results of my experiments would take away from the "detective interest" of the theoretical chapters.

view. Thus when a mother gives food to her child, we can see implicit in this cultural act a number of structural assumptions: that mothers feed children; that children are dependent on their mothers; that mothers are kind; that taro is edible; etc. etc. When we study a culture, we find at once that the same structural assumptions are present in large numbers of its details. Thus in the behaviour of all Iatmul mothers we can collect a series of diverse details, all of which imply that women look after children; and in a structural enquiry we may regard the phrase "look after" as a sort of shorthand by which we may refer to one structural aspect of all these details of cultural behaviour. Such shorthand is not the creation solely of the anthropologist; every culture contains generalised concepts which are a shorthand means of referring collectively to structural aspects of large numbers of details of standardised behaviour. In this structural approach, we shall see such phrases as "mother", "moiety", "patriliny", etc., as generalised abstractions for referring to the structural aspects of large masses of cultural details, i.e. details of standardised behaviour.

Used in this sense, the term, Structure, is closely related in meaning to the term "tradition". But the latter term has disadvantages; it is evidently out of place in any synchronic analysis, since in such an analysis the historical origin of any cultural detail is irrelevant. In spite of this the word "tradition" does sometimes creep into synchronic studies. In such contexts the word must clearly be stripped of all reference to the past or to diachronic processes. When so stripped, "tradition" can only mean the *given* facts of a culture, facts which are "given" as premises. Thus it comes about that the word "tradition" is sometimes used in synchronic analysis as a synonym for "cultural structure" as here defined.

The word, structure, is a convenient collective term for this aspect of a culture, but it has the disadvantage that it cannot conveniently be used to refer severally to the elements of which structure is built. I shall therefore use the word *premise* for these elements. Thus a premise is *a generalised statement of a particular assumption or implication recognisable in a number of details of cultural behaviour.*

We shall see that there is a great deal of variation in the extent to which the premises of a culture are explicitly stated by the people; that in many cases they may only be stated in symbolic terms—by the terminology of kinship, by metaphoric clichés, more obscurely by ritual, or, in Iatmul culture, by some trick in the giving of personal names. But in spite of this variation, it is convenient to use the term for any general statement whose expression can be directly collected from informants or which can be shown over and over again to be implicit in the behaviour of individuals.

The existence of premises or "formulations" in culture has, of course, always been recognised, but it is not sufficiently realised to what extent the various premises of a culture are built together into a coherent scheme. The student of cultural structure can detect syllogisms of the general type: "A mother gives food to children, a mother's brother is identified with the mother, therefore mother's brothers give food to children." And by these syllogisms the whole structure of a culture is netted together into a coherent whole. As I see it, the task of structural anthropology is the investigation of these schemes in culture, and I would define cultural structure as *a collective term for the coherent " logical"*[1] *scheme which may be constructed by the scientist, fitting together the various premises of the culture.*

Besides this use of the word, structure, to refer to an aspect of *culture*, there is a second perfectly valid use of the term to refer, not to culture, but to *society*. It is in this sense that the word is chiefly used by Radcliffe-Brown and it is important to keep the two senses distinct. In the study of cultural structure we take details of behaviour as our units and see them as linked together into a "logical" scheme; whereas in the study of social structure we shall take human

---

[1] In this definition I certainly am not using the term "logic" in that strict sense in which it is used in our own culture to describe a discipline of thought in which the species of steps involved are rigidly and consciously controlled. I do mean, however, that the elements of structure are linked together by steps. But it is probable that cultures may vary in the species of steps which link their premises together, and that the word "logic" must therefore be interpreted differently in every culture. This concept, the eidos, of a culture will be elaborated at a later stage in this work.

individuals as our units and see them as linked together into groups—e.g. as kin, as clan members or as members of a community. Such a difference as that drawn (p. 97) between peripheral and centripetal orientation of social systems might, for example, be described as a difference in *social* structure.

Thus in the two disciplines, the sociological and the cultural, we shall to a great extent be studying the same phenomena, but looking at them from two different points of view. In the study of cultural structure we shall see clans and kinship terminology as shorthand references to details of behaviour, while in the study of social structure we shall see these groupings as segments in the anatomy of the community, as a part of the mechanism by which the community is integrated and organised.

For the present I do not see my way to a clear isolation of the study of social function from that of social structure, and I believe that the most useful approach to these disciplines which we have at our disposal is that which sees social structure in relation to the integration and disintegration of the society. Therefore I shall consider these two subjects together under the general heading "Sociology of *naven*". But it is easy and, I believe, valuable to isolate the study of *cultural* structure from that of other aspects of the culture and these subjects I shall therefore treat separately.

FUNCTION

The term, Function, is more difficult to define since it is ambiguous. On the one hand we have the wide philosophical use of the term to cover the whole play of synchronic cause and effect within the culture, irrespective of any consideration of purpose or adaptation. Using the term in this sense we might say that the *naven* ceremonies are a function of all those elements of culture or properties of society upon whose presence the existence of the ceremonies depends. If, further, it can be shown that the *naven* ceremonies play some part in the integration of the community, we may say that the size of the Iatmul villages is a function of—amongst other things—

the *naven* ceremonies. This is a perfectly consistent and logical use of the term and one which anthropologists might do well to adopt.[1] They might then insert qualifying adjectives to indicate what sort of interdependence they are discussing. We should then have such phrases as: structural function, social function, pragmatic function, etc.

But unfortunately the word, function, may also be used in the popular sense to mean *useful adaptive effect* and it is often not perfectly clear in which of these senses the word is used in modern anthropology. Malinowski tends to define function in terms of adaptation, to regard all the elements of culture as "working directly or indirectly for the satisfaction of human needs", and he deduces from this that every detail of culture is "at work, functioning, active, efficient".[2] This method of approach is probably sound and its careful investigation might give a coherent system of anthropology allied to systems of economics based upon "calculating man". But unfortunately the other sense of the word, function, has confused the issue. Thus we find: "The simplest as well as the most elaborate artifact is defined by its function, the part that it plays within a system of human activities; it is defined by the ideas that are connected with it and by the values which surround it."[3] From this we must suppose that "function" is here being used not only in reference to the satisfaction of human needs but also in reference to the interdependence of elements of culture.

The position is complicated and difficult; and for me at least it is not simplified by the theory that culture may be subdivided into "institutions", where each institution is defined by its special function or functions. To me it seems premature to attempt such an analysis into institutions until the concepts covered by the term, function, have been analysed.

[1] The chief difficulty, which this use involves, is that on this principle the *naven* ceremonies could be shown to be a function of everything else in the universe and the anthropologist will be faced with the problem of defining his spheres of relevance. The attempt to subdivide culture into institutions is, no doubt, designed to delimit spheres of relevance.

[2] Both quotations are from the *Enc. Soc. Sci.* 1931, art. Culture, p. 625.

[3] *Loc. cit.* p. 626.

But if, instead of emphasising the subdivision of culture, we examine the possibilities of subdividing and classifying different types of functions, I believe that some degree of clarity can be achieved and that we shall then see some chance of unravelling the laws of social and cultural functioning.

A short and incomplete list of the chief kinds of effect to which the term, function, has been applied will suffice to demonstrate the present confusion and the sort of modification which I am suggesting:

1. The direct satisfaction of human needs.
2. The indirect satisfaction of human needs.
3. The modification, elaboration, etc., of human needs.
4. The moulding and training of human beings.
5. The integration of groups of human beings.
6. Various sorts of interdependence and relationship between elements of culture.
7. The maintenance of the *status quo*. Etc.[1]

In such a list the fact which chiefly strikes us is that every one of these types of function is dependent upon every other type. It is this fact which has so impressed the functionalists that they have fought shy of analysing the concept of function.

Scientific anthropology was born in an age when the older sciences had so far solved their domestic problems that they were beginning to extend their enquiries into the borderline areas which separate one science from another. The result of this extension has been a new sense of the fundamental unity of science and of the world. But the effects upon anthropology have been disastrous. The emphasis upon unity has retarded analysis. Physics and chemistry are benefiting greatly from a pooling of their problems and methods, but this does not mean that they did not benefit by their separation in the past. The great advances in knowledge are made by analysis of the problems, by the separation of one class of problem from another. To state that all problems are inter-related is mystical and unprofitable. To isolate them is artificial but

---

[1] The excellence of the field work done by members of the functional school is no doubt due to the width of the net which they throw into the sea of culture. I am not here advocating any reduction of this width but rather a greater discrimination in the sorting of the catch.

not more so than is the use of pure materials in a chemical laboratory.

The state of affairs in anthropology is perhaps not quite as bad as I have drawn it. We have on the one hand Radcliffe-Brown, who tends to see all the elements of culture in their bearing upon the solidarity, the integration, of the group. He insists that purely psychological considerations are irrelevant in this matter and he discriminates social structure from pragmatic function and regards each as presenting its special type of problem to be studied by its own special discipline. And on the other hand we have the developing science of economics which specialises in that type of social function which is concerned with the satisfaction of human needs and desires—a worthy methodological attempt perhaps hampered by a narrow view of human needs.[1]

These are considerable advances and in this book I shall build upon the disciplines discriminated by Radcliffe-Brown. I shall endeavour to consider the functional position (using the term in its widest philosophical sense) of the *naven* ceremonies from five different points of view. That is to say, I shall classify functions into five categories so that the various parts of the book will illustrate five different methods of approach to the problems of culture and society. As far as possible, I shall keep these methods of approach separate and shall indicate where I depart from this procedure.

The categories which I have chosen are:

1. *Structural* or "logical" relationships, between the cognitive aspects of the various details of cultural behaviour: the cognitive reasons for behaviour.

2. *Affective*[2] relationships, between details of cultural

---

[1] Sargent Florence (*Economics and Human Behaviour*, chap. IV, 1927) discusses the position taken up by Henderson in economics—the assumption that we can retain "calculating man" as a basic postulate of economics provided we make no assumptions as to the terms in which the calculation takes place. This restriction provides a convenient methodological isolation of economics from psychology.

[2] Although the terms "affect" and "cognition" are now in some disfavour among psychologists, I have been compelled to adopt the adjectives derived from them. I believe that the dangers and fallacies inherent in these terms may to a great extent be avoided if we speak not of "affective mechanisms" and "cognitive mechanisms" but of affective and cognitive *aspects* of the behaviour of a single mechanism, the individual.

behaviour and the basic or derived emotional needs and desires of the individuals: the affective motivation of details of behaviour.

3. *Ethological* relationships, between the emotional aspects of details of cultural behaviour and the emotional emphases of the culture as a whole.

4. *Eidological* relationships, between the cognitive aspects of details of cultural behaviour and the general patterning of the cultural structure.

5. *Sociological* relationships, between the cultural behaviour of the individuals and the needs of the group as a whole: the maintenance of solidarity, etc.

Of these five categories of relationship between bits of anthropological material, it is clear that one, the sociological, is in a class apart. Of the others, the ethological is very closely related to the affective, and the eidological is very closely related in an analogous manner to the structural.[1]

Besides these, there are two other methods of approach which I shall not consider in any detail, namely, those of developmental psychology and economics. These are omitted, not because I decry their importance but because I did not collect in the field the material which would have illustrated these approaches.

Of these categories but little need be said at this stage in our enquiry. The concepts of structure, both social and cultural, I owe very largely to Radcliffe-Brown; while another concept, that of "pragmatic function", was derived from Malinowski. This latter concept I have subdivided for several reasons.

Let us consider some rather more concrete examples of "pragmatic" functioning. The term is applied to such effects as: increase of sociability among individuals; increased solidarity of the community; increased family pride; the confirmation of the privileges of individuals and the enforcing of their rights and duties; substantiation of belief in magical

---

[1] Indeed so close are the resemblances between these pairs of categories that I have been led to refer to the affective functions of behaviour as "the expression of ethos in behaviour"; and I might similarly have referred to the structural functions as "the expression of eidos in behaviour".

efficiency and the strengthening of traditional law and order.[1]
At a first glance many of these functions would appear to be
"useful", and anthropologists have been tempted to attach
value to functions of this sort without more ado. But their
usefulness becomes more problematical when we realise that
many of these effects may be mutually antagonistic, that for
example an increase in family pride may well disrupt the
solidarity of the community; or, again, excessive emphasis
upon solidarity within the community may conceivably lead
to wars with outside peoples and so to the destruction of the
status quo. In the light of the potential antagonism between
these effects, it would seem that value can only be attributed
to them with great caution; and in a sample list of valuable
effects we should certainly include all the opposites of the
effects stressed by the Functional School—the reduction of
solidarity, the reduction of family pride, the relaxing of
duties, etc.

Again, if we consider this sample list of functions in diverse
settings, their value becomes still more problematic. A nomad
people living on the barest margin of physical sustenance,
and drastically regulating population by infanticide, is not
likely to be interested in integrating a larger population. The
truth of the matter is that different peoples attach very
different values to these various effects. Some peoples approve of sociability while others are inclined to frown upon
the sociable talkative individual and to regard him as un-
dignified and intruding. The "strength of traditional law"
is very important in some African communities, but the
phrase is almost meaningless when applied to the Iatmul who
have a highly individualistic culture and will readily respect
the law-breaker if he have but sufficient force of personality.
The Iatmul too—like the Irish—set no exaggerated value
upon extremely orderly behaviour. They desire a degree of
orderliness very different from that demanded by, say, a
Quaker community. And so on; whichever of these effects
we take we shall find that among certain peoples that effect

[1] This sample list of functions, and the term "pragmatic function"
applied to them, are derived from Malinowski's article "Anthropology",
Supplement to the *Encyclopaedia Britannica*, 1926. They are the functions
which he ascribes to myths of various types.

is highly valued while among others it is ignored or despised. The same conclusions apply not only to such abstract effects as the strength of law and integration, but even to such concrete matters as the food supply, one people setting a high value on good living while an ascetic group may take but little interest in the flesh pots.

It is, I think, time that anthropologists took account of this enormous variation in the value which different peoples set upon the effects which modern functional anthropology has to offer them. In the light of this variation it is obviously dangerous to state that the significant pragmatic function of a given detail of a given culture is the increase of family pride—unless it can first be shown that family pride is really one of the effects which is valued in other contexts of the particular culture which we are studying.

Here, then, we arrive at a very different type of functional relationship. We began by considering the relation between details of culture on the one hand and the needs and desires of the individual on the other. But we are now faced with the fact that we cannot guess at those needs and desires but must first deduce them from the emphases of the culture as a whole. Thus, if we isolate from "pragmatic function" the concept of *affective function* which we may define, rigidly, in terms of the relation between details of culture and the emotional needs of individuals, it follows that we must construct another category for the relationship between the emotional content of the particular detail of behaviour whose functions we are studying and the emotional emphases of the culture as a whole. This category of function I shall call *ethological*; and I shall use the word *ethos* to refer collectively to the emotional emphases of the culture.

Similar considerations (see footnote to the word "logical", p. 25) will lead us from the isolation of the concept of cultural structure to the isolation of *eidos*. From the examination of the premises in the structure of a particular culture we can fit them together into a coherent system and finally arrive at some general picture of the cognitive processes involved. This general picture I shall call the *eidos* of the culture.

But, though the relationship between eidos and structural

premises is so closely analogous to that between ethos and affective functions, there is one anomaly to which attention must be called. We have seen that the study of ethos is a necessary preliminary to any conclusions about pragmatic functions. But in the case of structure and eidos, the order of procedure is reversed. The details of the cultural structure are to be first studied and from these the eidos is deduced. This reversal is due to the fact that in studying cultural structure we are concerned with the manifest cognitive content of behaviour, whereas in studying pragmatic functions we are concerned with the much more obscure emotional content. The manifest content can be *described* piece by piece and the underlying system deduced from the resulting description. But emotional significance can only be *ascribed* after the culture as a whole has been examined.

Apart from this difference in procedure, the ethological and eidological approaches to culture are very closely analogous. Both are based upon the same fundamental double hypothesis: that the individuals in a community are standardised by their culture; while the pervading general characteristics of a culture, those characteristics which may be recognised over and over again in its most diverse contexts, are an expression of this standardisation. This hypothesis is, in a sense, circular; it is supposed that the pervading characteristics of the culture not only express, but also promote the standardisation of the individuals.

For this double assumption, we are chiefly indebted to Dr Benedict; and the concepts, ethos and eidos, which I am suggesting may be regarded as subdivisions of her more general concept, *Configuration*. The eidos of a culture is an expression of the standardised cognitive aspects of the individuals, while the ethos is the corresponding expression of their standardised affective aspects. The sum of ethos and eidos, *plus* such general characteristics of a culture as may be due to other types of standardisation, together make up the configuration.[1]

[1] Dr Benedict was kind enough to read a large part of this book in manuscript form before it was completed and she agrees with me that this is a correct statement of the relationship of ethos and eidos to configuration.

This hypothesis of the standardisation of the individuals goes a long way to make clear the inter-relationships of the various disciplines which we have so far isolated—the structural and eidological, pragmatic and ethological—but it also brings to the fore another separable discipline, another category of function. This discipline is generally called Developmental Psychology, the study of the moulding of the individual by the circumstances of his environment, the psychological effects upon him of the impact of culture.

Thus we have worked down from the study of cultural processes to the study of individual psychology, linking the two ends of the scale by the abstract[1] concept of the "standardised individual".

Returning now to the raw material of anthropology, the facts of native behaviour noted down by the field worker, we may see the whole of this mass as relevant to a final separable discipline. This is concerned with the effects of behaviour in satisfying the needs of *groups* of individuals, in maintaining or impeding the continued existence of that abstract organism—the Society. This category of function I shall call *Social Function* and shall use the term *Sociology* in a restricted sense to refer to its study, as distinct from that of the other separable disciplines.[2] In this discipline of sociology, as Radcliffe-Brown has pointed out, individual psychology becomes irrelevant to the same extent that Atomic Physics is methodologically irrelevant in the study of Biochemistry.

[1] It may be possible in the light of future work to define the "standardised individual" in rather more concrete terms. For the present it is not possible to say whether by this concept we mean the "average" or the "ideal" individual. Nor is it at all clear what differences will be introduced into the theory of culture according to our interpretation of the "standard". These matters can only be cleared up after more intensive work has been done on normal and maladjusted individuals in various cultures.

[2] Malinowski, I believe, includes under the term "pragmatic function" not only the type of function to which I apply that term but also what I am here calling "social function". The difference in our points of view springs largely from his definition of function in terms of the direct *or indirect* satisfaction of the individuals. I have defined pragmatic function in terms of *direct* satisfaction only, leaving the various steps in the roundabout routes of indirect satisfaction for separate examination in the other categories of function.

CHAPTER IV

Cultural Premises relevant to the
Wau-laua *Relationship*

IN this section and that which follows it, I shall view the elements of Iatmul culture as links in a structural system. I shall see the mother's gifts of food to her child and the child's loyalty to its mother as details of behaviour which are assumed and expected in this culture, as premises. Later, in another section, I shall consider such cultural facts in their relation to needs and desires. I shall for the present endeavour only to show that the premises of Iatmul culture are connected together into a coherent, "logical" system. I shall consider syntheses of the type: a mother gives food to her child. A mother's brother is identified with the mother. Therefore a mother's brother gives food to his sister's son.

For the handling of cultural elements as parts of a structure Radcliffe-Brown has suggested a very useful technique involving the use of the technical term *identification*. This term has given rise to a certain amount of misunderstanding and I must explain the sense in which I use it. The term is a shorthand method of referring collectively to the structural aspects of large numbers of details of a culture. When I say that siblings are *identified* I mean that in this culture there is a large number of details of culturally standardised behaviour in which the behaviour of one sibling resembles that of the other; and I also mean that many details of the behaviour of outside individuals towards one sibling are reproduced in their behaviour towards the other. Further, we shall see that, in Iatmul culture at least, when two individuals are identified it is usual for one of them to behave as if he had performed certain acts which as a matter of fact were performed by the other.[1] Similarly, outside persons behave

[1] The most striking case of this occurs in the identification between living individuals and their ancestors. Among the Iatmul one often hears such statements as "*I* was on the stern of the canoe which brought the first people to Mindimbit village!", meaning that an ancestor who had the same name as the speaker played this part in the founding of the village.

towards the one individual as if he had performed the acts which were performed by the other.

The statement that siblings are identified does not imply the absence of details of behaviour which would discriminate between them. Thus the word identification is used relatively and not absolutely. It is common to find that in one series of contexts two individuals are identified, while in another series they are contrasted. Thus we shall see that, in Iatmul culture, father and son are identified in their relations with the outside world while between themselves they are drastically contrasted. In so far as they are identified, they are regarded as allies but in so far as they are contrasted, they are regarded as mutually opposed in their interests.

The technique of describing identification is by far the best so far evolved for the mapping of kinship systems; and for the study of *naven* we have to consider one important discrimination and four major types of identification:

1. That mother's brother (*wau*) is distinct from father's brother or father (*nyai'*). This discrimination is made overtly in the kinship terms and is implicit in many of the patterns of behaviour towards these relatives. In the present context this premise may be suitably illustrated by a myth about *naven*:

A boy and girl—brother and sister—went out to work sago in a sago swamp. As they went in their canoe the mayflies came out and the surface of the river was covered with them, so they set to work and collected a great quantity of mayflies. The fish too began to rise and the boy speared a number of fish. Then they went on to the sago swamp where the boy felled a sago palm and beat the pith out of the trunk. The girl took the pith and washed it and collected the sago. On their way home they were met by their father and mother who had come out to meet them in a canoe, and who exclaimed: "Who speared the fish?" and "Who beat the sago?" The boy and girl replied that they two had speared the fish and beaten the sago. Then the father took off his pubic apron and the mother took off her skirt and they began to dance a *naven*. The boy and girl exclaimed: "Why are you dancing?" and were utterly ashamed. So the boy and girl put the sago and the fish back in their canoe and paddled away to their *wau*. When they reached the house, *wau* and *mbora* (mother's brother's wife) came out and took up the fish and sago, and *wau* took off his apron and *mbora* took off her skirt and danced a *naven*, and *mbora* took up

the fish and sago and cooked them and she and *wau* ate. Later
mbora brought a pig which she had fattened and *wau* presented it
to the two *laua* who set their feet upon it. Then they went and
prepared shell valuables and presented them to the *wau*.

This discrimination between father and mother's brother
is especially conspicuous in all those contexts in which their
behaviour is influenced by their membership in different
patrilineal clans. Every clan has its series of myths about
ancestors. These myths and the spells which are based upon
them are secrets which are carefully preserved by the descen-
dant clans. A man will normally pass on his esoteric knowledge to one of his sons; but sometimes he will impart the
clan secrets and spells to a sister's son. In the latter case the
esoterica *must be ceremonially paid for* with shell valuables
and the sister's son is supposed not to pass on the *wau*'s
esoterica to his own sons.

The entrusting of esoterica to the *laua* is a strong affirmation of the tie between him and his *wau*. It implies considerable trust and confidence on the *wau*'s part since the
laua, when he has been told the secrets, may easily use them
in debate to his *wau*'s undoing and shame. The motive which
leads the *wau* to take this risk is, I think, a hope that the *laua*
will speak on his side in debates. Malikindjin, who spurned
the idea that he needed any assistance in debate (cf. p. 165),
yet taught a part of his esoteric learning to his *laua*, saying,
"When I am dead, my *laua* will help my son in his debates".
I was told of one case only in which a *laua* used, in debate
against his *wau*, information which the latter had confided
to him.

Sometimes a clan has the misfortune to lose its ancestral
traditions and must then *buy* them back from its *laua*s.

An interesting case of discrimination between *wau* and
father occurs in Mindimbit,[1] in the procedure which accompanies a small boy's first killing of a human being. For this
important occasion the father of the boy will obtain, either
by capture or purchase, a captive from a neighbouring com-

---

[1] This ceremonial point is apparently not made in Kankanamun and
Palimbai, where I asked my informants about the corresponding procedure. They seemed to think it did not matter who helped the boy in
his kill: that probably the father would do it.

munity and this captive will be speared by the boy. But the killing is in many cases carried out while the child is still too young to wield a spear; and so he must be helped in the performance of the deed. This help must not be provided by the father because the natives say that the kill would in that case be the father's achievement and not that of the son. But the *wau* can help the child; he will hold the spear and direct the child's thrust, and after this the kill will count as the achievement of the child.

Similarly *wau* is discriminated from father in economic contexts. Here the pattern is the inverse of that in contexts of achievement. When a boy is collecting valuables to provide a bride price, his father will, as a matter of course, help him and will not expect any formal recompense for the valuables which he contributes. The *wau* may also contribute, but in this case he will expect some return. Usually the boy will make this return in the form of a contribution to the *wau*'s son when the latter is collecting valuables to get a wife. Thus, in contexts which concern valuables, the boy is grouped with his father and separated from the maternal clan; while in contexts of achievement the boy is united with the maternal clan and separated from his father. But in both types of context the *wau* is discriminated from the father.

In spite of this discrimination, we shall see that there is also a trace of identification between father and *wau* and that this appears even in some details of their behaviour towards the *laua* (cf. p. 83). In other contexts, in which the same two persons are concerned, not dealing with the *laua* to whom they appear as *wau* and *nyaï*, but dealing with the outside world to whom they appear as a pair of brothers-in-law, we shall see that co-operation and competition, identification and discrimination, are very evenly mixed in the patterning of their mutual relationship.

2. That there is a certain identity between a father and his son. This premise is expressed, for example, in the patrilineal system of clans, and in many other ways. For the present purpose it will suffice if we examine in some detail only one of the contributory lines of evidence—that from the vocabulary of kinship—and omit the others. I have

already published[1] a series of examples of the Iatmul tendency to summarise and formulate the behaviour of individuals by using the terminology of kinship in an analogic manner. This usage is sufficiently familiar in our own society to need no stressing, and since the Iatmul have carried this practice further than we have, we can with unusual safety refer to the vocabulary as evidence of the formulated structure of the kinship system.

The vocabulary of kinship contains a number of phrases in which two kinship terms are used in juxtaposition to indicate a group of relatives who are closely associated in the native mind, and from such compound forms we may deduce a close identification between the behaviour of and towards the types of kin who are actually named in the compound phrases. As an example we may take the term *nggwail-warangka* in which the word *nggwail* means father's father and *warangka* means father's father's father. The whole compound is a collective term for patrilineal ancestors, a group of relatives who are classified together in Iatmul culture.[2]

There are several other cases in which compound terms meaning "a certain relative and his father" are used to denote whole groups of relatives who are identified together, for example: *wau-mbuambo* (mother's brother and mother's brother's son), which is a collective term for all the members of the maternal clan; *towa-naisagut* (abbreviated from *tawontu-naisagut*, and meaning literally wife's brother and wife's father), which is a collective term for the whole of the wife's clan; *laua-ianan* (sister's son and sister's son's son), a collective term for the descendants of the men who married the speaker's sister. In all these terms the identity of a son with his father is emphasised.

A similar identification is implied by an incorrect use of kinship terms for the mother's brother's son and mother's brother's son's wife. These two relatives are normally spoken of as *mbuambo*, a term which identifies them with mother's father and mother's mother. But one informant in Palimbai

---

[1] *Oceania*, 1932, p. 266.
[2] The same set of relatives is constantly invoked in name-songs, in the refrain: "*nyai' nya! nyamun a!*"—(O Father! O Elder brother!).

told me that after the death of the mother's brother, the term *wau* is used for the mother's brother's son, and the term *mbora* (mother's brother's wife) is then applied to the mother's brother's son's wife. Other informants said that this was incorrect, but I suspect that if the culture were left undisturbed this usage might become more general. Among the neighbouring Tshuosh people there is no distinction in terminology between the male members of different generations of the maternal clan; all are called *wau*. Among the Iatmul too, such a use of terms would fit with the *wau*'s motive in teaching his esoterica to his *laua*—"When I am dead my *laua* will help my sons in their debates" (cf. p. 37).

Here it must be stressed that the identification of father and son, while it is conspicuous in their relations with outside persons, is not allowed to become a basis for easy intimacy between them. Such intimacy is sternly discouraged, and in its place the conflict between successive generations is emphasised and is handled in a dignified manner. While the father may resent bitterly and avenge any scolding or blows which others may administer to his son, he himself is in a disciplinary position and may strike his son in anger. But in contrast to the disciplinary position of the father, we find that the son will gradually displace the father. These two facts are the basis of the formulations which govern their relationship. The possibility of any overlapping between father and son, in social status, is drastically limited by taboos which are felt not to be arbitrary but rather to indicate the only proper way of meeting the situation. The father must never profit at the expense of his son. He will never eat anything which the son has grown in his garden, nor may he eat food gathered by his son, even though it be from the father's own garden. When they are in a canoe together, the son normally takes the position of honour in the bow and, while he is there, the father will never scoop up water from alongside to quench his thirst. In the initiatory system the son is normally two grades below his father (see diagram, p. 245). But the people are continually readjusting the balance between the grades and it occasionally happens that an attempt is made to promote a boy into the grade of which his father is a member. Such an attempt is violently resisted by the boy, who feels that this will be to intrude upon his father's dignity. But the father will make no resistance to his son's promotion, since it ill becomes him to defend his own pride at the price of his son's advancement. I think that in most cases the promotion is finally effected. But in one case which I witnessed it was successfully resisted by the son after he had finally said that he would abide by his *wau*'s decision. The *wau* was consulted and opposed the promotion. Thus the son is as much concerned as the father to avoid any overlapping of social status.

The taboos which a father observes are obeyed chiefly because the father feels them to be right. But they are to some extent reinforced by a belief that if he eats food collected by his son he will soon grow old—perhaps a symbolic statement of the fact that in so doing the father will be behaving like a grandfather.[1] But more important than this sanction is the strength of public feeling. The idea of intimacy between father and grown son is shocking to the Iatmul. While I was on the Sepik, a European father and son settled together on the Ceram River many miles away, where they started a small, isolated tobacco plantation. The Iatmul heard of this and were surprised and shocked. They came to me to ask if the two were really father and son and were disgusted when I told them that it was so. They said, "Has the father no shame?"

These taboos have very little classificatory spread. They apply only to own father, father's own younger brothers (father's elder brother is addressed as *mbuambo*, a term which means, among other things, "maternal grandfather") and own father's *tambinyen* or partner in the opposite moiety.[2]

The behaviour patterns between father and son are marked

---

[1] In Iatmul social organisation, alternate generations in the patrilineal line are grouped into two opposite *mbapma* (literally, "lines") as in Australian kinship systems. It follows from this that a father, in identifying himself with his son, would be behaving as if he were the boy's grandfather.

[2] See *Oceania*, 1932, p. 264.

typically by restraint on both sides. Each respects the other, and both are rather quiet and stiff when talking together. But this does not mean that an outsider who is speaking in the presence of one of them can make remarks about the other. A very real identification of interests and of needs exists between them in spite of their mutual avoidance of overlapping status.

3. That in spite of the patrilineal system, the child is linked in some way with its mother's clan. This premise may be recognised, symbolically expressed, in the native theory of gestation. It is supposed that the bones of the child are a product of the father's semen, while its flesh and blood (somewhat less important) are provided by the mother's menstrual blood. This idea is carried logically to the conclusion that the afterbirth, lacking bones, is therefore the child of the mother only.

The child is in one way a member of its father's clan, but at the same time it is in quite a different way a member of its mother's clan. This formulation is expressed by the giving of two sets of names to the child. One set are names of the totemic ancestors of the father's clan, while the other set are names of the totemic ancestors of the mother's clan. These two sets of names have different terminations: the patrilineal names end in suffixes which mean "man", "woman", "body", etc., while the names of the ancestors of the mother's clan end in the suffix, -*awan*, which probably means "mask" and is perhaps connected with the custom mentioned above of dancing in masks which represent the totemic ancestors of the maternal clan.[1]

These two sets of names are apparently felt to represent two quite different facets of the individual's personality. After death a man dwells in the land of the dead under the names which his *wau* gave him, but he is, at the same time, reincarnated in his son's son under his patrilineal name. In the techniques of black magic it is the -*awan* name which is

---

[1] The etymology of suffixes attached to the names which are given by the *wau* is doubtful. The corresponding names given to female children end in the suffix, -*yelishi*. This word means "old woman" or "old lady". It is possible that the word *awan* means "old man" and that this word is only applied metaphorically to the masks called *awan*.

used by the sorcerer in addressing the figure which represents his victim; and the people take some slight care not to use the *-awan* names after dark, lest the witches overhear and use the name to the detriment of its owner.[1]

In general, it would seem that the names given by the *wau* represent a more mysterious aspect of the personality than those given by the father. In a sense the link with the maternal clan is analogous to that with the paternal and each is expressed by the giving of names; but while according to the patrilineal name the child is a "man" or "body", according to the matrilineal he is a "mask". Similarly in the patrilineal theory of reincarnation the child is *concretely* stated to be a reincarnation of a father's father, while his identity with the ancestors of his mother's clan is only *symbolically* expressed in ritual and ceremonial behaviour. This ritual behaviour may conveniently be enumerated at this point:

(*a*) The giving of *-awan* and *-bandi* names.

(*b*) The use of the term, *nyai' nggwail*. This term means literally "father and father's father", but it is not used as a term of address within the clan except in invoking the ancestors collectively. It is, however, the ordinary term of address used reciprocally between *wau* and *laua*. The term which is correspondingly used between members of the same patrilineal group is *bandi*, which also means "novice", "initiate" or "young man".[2]

(*c*) The custom of *tshat kundi*. "*Nyai' nggwail*" is the ordinary form of address used by the *wau* in talking to his

---

[1] One informant in Palimbai told me that the name used in sorcery is not the *-awan* name, but the *-bandi* name. The word *bandi* means an initiate (i.e. either a novice or an initiated person) and names with this suffix are given to novices by their *waus* at the end of initiation. The same informant said that women and children who have no *-bandi* names could therefore not be killed by this species of magic. The *-bandi* name, like the *-awan* name, refers to ancestors of the maternal clan.

[2] There is a curious tangle here, which I have not been able to unravel. In the light of the general identification between *laua* and *ancestors* it is understandable that the *wau* should call his *laua* "father and father's father". But it is not at all clear why the *laua* should use the same term for the *wau*. Still less is it clear why brothers and fathers should address each other by the term *bandi*, seeing that the *-bandi* name is given by the *wau*. The term, *bandi*, is definitely more intimate than *nyai' nggwail*.

laua, but on more marked occasions, when he is greeting him, hailing him, congratulating him or bidding him farewell, the *wau* will use more elaborate forms of words. These are known as *tshat kundi*.[1] In its simplest forms, used, e.g., when a *wau* is saying goodnight to his *laua*, *tshat kundi* consists simply in addressing the *laua* as a pair of objects which are ancestors of the *wau*'s clan. The *wau* may say—"You stay, coconut and areca palm, I am going to sleep." In this phrase the palms, by the everyday words for which the *laua* is addressed, are totemic ancestors of the *wau*'s clan.

On more important occasions the *wau* will use not a pair of everyday words for the totemic objects but a whole string of the personal names of some important totemic object or ancestral hero. He will greet his *laua* from a distance as he enters the village, shouting to him:

"Come, Father and Father's father, Tepmeaman and Kambuguli, Weimandemi and Tanggulindemi, Ulakavi and Tshugukavi, Weiuli and Weikama, Tshugutshugu and Nggaknggak, come, Father and Father's father. You are my *laua*."[2]

In this speech the names are those of the Borassus palm, which is also a fish (cf. p. 234) and an ancestor of the Mwailambu clan. A *wau* who is a member of this clan might use this series of names in hailing his *laua*, but there are also many other strings of names of important ancestors of this clan, and any of these he might use for the same purpose. He might even use several strings of names one after the other. Members of other clans would use the strings of names of their special ancestors.

Most of the names which are used in *tshat kundi* might also be given to sons of the *wau*'s clan. But there are also a few strings of names which have apparently dropped out

[1] *Tshat* is a verb which means to "*step across*" something, as, e.g. in the *naven* ceremonial, the hero "steps across" the women. *Kundi* is the ordinary word for "mouth", "speech", "language", etc. There is another phrase for the same custom—*tshivera kundi*—but I do not know the literal translation of this.

[2] In this speech, I have inserted the word "and" between the names of each pair. In the declaiming of the speech, the intonation would give the pairing and no copula would be used.

of use in the naming of children and are now only used for *tshat kundi*.

Sometimes *tshat kundi* is used by persons other than the *wau*, e.g. as a device for emphasising the fact that the person addressed is a "brother" of the speaker in the sense that both are *laua*s of the same clan. In all cases the names used are those of ancestors of the maternal clan of the person addressed.

Lastly, the names used in *tshat kundi* are chosen merely because they refer to important ancestors of the maternal clan. The choice is not guided in any way by the particular *-awan* or *-bandi* names which have been given to the person addressed—except in so far as the *-awan* name indicates to what clan the person's mother (or classificatory mother) belongs.

Thus we may sum up the custom of *tshat kundi* by saying that the *laua* is ceremonially addressed as if he were a random collection of his mother's important ancestors.

(*d*) The *laua* dances in masks which represent the ancestors of the maternal clan. He blows the flutes which are their ancestors. He carves representations of their ancestors on the house-posts which are to stand in their ceremonial house. He carves the ancestral *mwai* heads (cf. Plate XXVIII B) which are to adorn the masks of the maternal clan—the masks in which he will later dance.

(*e*) The *laua* is entitled to ornament his body with those plants which are totemic ancestors of his maternal clan; and he will resent the use of these plants by persons not so entitled.

(*f*) When the members of a clan sacrifice to their clan ancestors, it is the *laua*s who eat the sacrifice.

A similar position is brought out with diagrammatic clearness in the procedure in cases of offence against the rights of land tenure.

The land of a clan is watched over by the *angk-au*. This word means literally "potsherds" and is a term for those spirits of ancestors which are symbolised by the old broken sherds which lie under the house, sherds which have been thrown away by generations of previous inhabitants of the house. When the *angk-au* see any outside person violating the land of the clan, e.g. planting upon it or stealing its products, they make a little noise in the house at night, "like a rat squeaking". The owner of the house hears this sound and goes out next day to see what is amiss. He finds that somebody has planted coconuts on his land. Then he puts *ndjambwia* on the ground. *Ndjambwia* may be spikes set up to wound the foot of the trespasser, or they may be magical objects to cause sickness of the trespasser; the same term is used for both techniques. Later the trespasser comes and his feet are wounded and develop sores. The *angk-au* will prevent the healing of these sores till the offender goes to the members of the clan who own the land. Then the offender will kill a fowl and present coconuts to them. He will also go to the patch of ground and will there hang up a basket of valuables and a *tambointsha*.[1] The *angk-au* will take the "soul" of the valuables and of the *tambointsha*; and after a few days the valuables will be returned to the trespasser. But the fowl and coconuts will be eaten by the *laua*s of the clan which owned the ground.

(*g*) The *laua* eats his maternal ancestors. On one of the lakes in the neighbourhood of Mindimbit village there is an island of floating grass or sud (*agwi*); and on this a certain crocodile lays its eggs. This crocodile has the personal name, Mwandi-ntshin, and is an ancestor of the Mandali clan. Members of this clan go from the village to collect the eggs. It is said that they find as many as sixty eggs and among them two double eggs. The ordinary eggs they eat but the double ones are presented to their *laua*s, who eat them.

In this detail we have an isolated case of the *laua* eating *wau*'s ancestors. But besides this there is also a regular ritual in which the same act is performed. This ritual is called *pwivu*. The maternal clan prepare in a large bowl a mixture containing scrapings from the bones of their ancestors

---

[1] *Tambointsha* are tassels of feathers tied to string. They are symbols of successful homicide and worn on the lime stick as a tally of the owner's successful kills. A kill may be scored for homicide achieved by means of *ndjambwia*, but not for homicides achieved by other magical techniques. The act of presenting a *tambointsha* to the *angk-au* thus falls into the same pattern as the presentation of *tambointsha* described on p. 99. In both cases we may see a formal recognition of the fact that the damage done was a legitimate act of vengeance.

and other curious clan relics. This mixture is eaten by the children of the village—but especially by the *laua*s of the clan whose ancestors have gone into the mixture.[1]

(*h*) Finally, when the *laua* dies a ceremony called *min-tshanggu* is performed. A figure, which has for head the modelled and decorated skull of the *laua*, is set up on a platform which is suspended from the roof of the house so that at a push it will swing to and fro. This platform is called the *agwi* (floating grass island) and the spectacle represents the voyage of the ghost, on a drifting patch of floating grass, down the Sepik River to the land of the dead. The platform is suspended about two feet from the floor and from its edge hangs down a fringe of strips of palm leaves, forming a screen so that the women in the house cannot see into the space under the platform. This fringe is called *tshimbwora*, the word which is used for water in the special ritual jargon of the shamanic spirits.

The ritual which accompanies this spectacle is divided into two halves, the first half being performed by the members of the dead man's clan, and the second half by members of his mother's clan. The two halves are essentially similar, each consisting in the singing of name songs. In the particular celebration which I witnessed, the members of the dead man's clan collected together in front of the *agwi* soon after dusk and sang name songs till after dawn next morning. Their clan flutes had been smuggled into the house and hidden under the *agwi*. From time to time the succession of name songs was varied by music from their flutes played by men (their *laua*s) hidden under the *agwi*. The same hidden men also occasionally caused the suspended *agwi* to swing, representing the rocking caused by waves in the water.

Early next morning the members of the dead man's clan dispersed and in their place the members of his mother's clan assembled. They brought with them a number of totemic

[1] A more detailed account of the *pwivu* ritual is published in *Oceania*, 1932, p. 472. Among the Iatmul the eating of the maternal clan is only found in these somewhat symbolic forms. But Iatmul informants told me that among a neighbouring tribe, on the Sud River, it is usual for the sister's child to eat a piece of flesh from the thigh of his mother's dead brother.

objects, their ancestors—branches of trees, paddles, etc. These they added to the decorations of the *agwi*, which had already been decorated with the totemic emblems of the dead man's clan. They also brought a hooked stick and their own clan flutes which were substituted for the others under the *agwi*. Then they began to sing their name songs and continued in this until late afternoon. During the singing, the hooked stick was used from time to time to swing the *agwi*: the members of the maternal clan overtly pulling the *agwi* and the figure of the dead towards themselves.

The three premises which I have mentioned deal with the relationships of the child to its father, mother and maternal clan. These may be summed up in a form especially relevant to *naven* by saying that the child is closely identified with its father but competes with him. The child's identity with its mother and its link with the maternal clan are more obscure. But here the child is not a competitor but, rather, an achievement of the mother; and the child's achievements are her achievements, the triumphs of *her* clan.

This conclusion can be further documented from another mortuary ritual. When a great man dies a figure is set up by the members of his initiatory moiety to represent him and is decorated with symbols of all his achievements. Spears are set up to the number of his kills and baskets are suspended from the shoulder of the figure to the number of his wives. But no symbols are added to show how many children he has begotten. For a woman, when she dies, a post (*nggambut*) is set up and is decorated with the ornaments which were given to her sons by the senior age grade at initiation. Her greatness is vicarious and lies in the achievements of her sons.

Similarly in the list of occasions which are celebrated by *naven*, we find that *naven* are celebrated for a boy when he takes a wife, but not when his first son is born. For a girl *naven* are celebrated when she gives birth to a child, but *not* for her marriage.

There is a curious myth which is probably relevant in this connection, as showing what might happen if the *naven* patterns were disregarded. In the myth *naven* are celebrated

for a man by his wife's brother and wife's father on the
occasions of the birth of both his first and second children.
These relatives present *valuables* to the husband and rub
their buttocks on his shin. He in return presents *food* to
them (i.e. the normal pattern of presentations is reversed).
After the second *naven* the relatives stay and stay demanding
more food, until the husband says, "There is none left."
Then they say, "You are lying." The husband said, "I am
not lying", and when he had said that, he would have speared
the child. But his wife—it was she who speared the child.

With this clue to the general sense of these formulations,
the procedure at the little boy's first kill falls into place as
consistent with the general pattern. The boy's triumph is
the triumph of his maternal clan; therefore it is allowable
for his *wau* to help him in the holding of the spear. But if
this office were performed by the father we may suppose that
the kill would be claimed and celebrated by the father's
maternal clan.

Lastly, we may enquire into an apparent discrepancy in
the premises which pattern the relationship of a child to its
mother's clan. On the one hand we have details of behaviour
which indicate that the child is the achievement of the
mother and its achievements are her achievements, and on
the other hand we have details which indicate that the child
is the ancestors of its mother's clan. This, I believe, is a
problem which cannot be satisfactorily solved until we have
examined the ethos of Iatmul culture. In later chapters, the
association between pride and ancestors will be documented;
and I will here overstep the strict limits of structural enquiry
by suggesting that *pride* is the factor which links together
achievement and ancestors, two apparently different ideas,
alike only in their emotional content.

4. That there is an element of identity between a brother
and his sister. This premise like the last is shown diagrammatically in the system of personal names. Siblings of
opposite sex are frequently given the same name with differences only in gender termination (cf. p. 243). The same
formulation is recognisable in the formation of a compound
kinship term of the type to which I have already referred.

This is the term *wau-nyame* (mother's brother and mother), which is used as a collective term for the maternal clan, as a synonym in fact of the term *wau-mbuambo* to which I referred above. The same identification is also indicated by the terms used between brothers-in-law. These terms are *lan-ndo* for sister's husband, and *tawontu* (i.e. *tagwa-ndo*) for wife's brother. The term *lan* means husband and the term *tagwa* means wife. Thus the sister's husband is called "husband-man", the male speaker identifying himself with his own sister in their relationship to her husband; while reciprocally, the wife's brother is called "wife-man", the speaker identifying wife's brother with wife. Both these terms therefore imply the identification of the brother and sister. A concrete example of the working of this identification may be seen in the myth of the brother and sister who went together to work sago, quoted above.

In this identification of siblings of opposite sex there is a type of insistence of which we shall meet other cases later, an insistence that the mere genealogical position is not enough; the identification must be made real in behaviour— more especially in ceremonial behaviour. In the passing on of clan names, which are given on the basis of the patrilineal alternation of generations, a man has a right to give the names of his classificatory and own fathers to his sons. But the women of the clan also have clan names and a man cannot simply take his father's deceased sister's names and give them to his daughters.[1] Before this can happen the identification of a brother with the woman must apparently be demonstrated in ceremonial behaviour. This is done either immediately after the woman's death or after the death of her husband or her brother:

[1] This would seem to be the general sense of the statements which I collected about this matter: one informant, however, stated that not only the female names but also the male names must be got from the father's sister. The case which we were discussing was one in which a man on the day after his father's death went and cut his father's sister's hair, "buying both his father's and her name" with a *Turbo* shell.

I do not know the meaning of the symbolic water into which the *Turbo* shell is put; it may perhaps be a reference to the mythological origin of these shells from the waves of the sea and a genital symbol. This piece of ceremonial is the only context I know of in which valuables are *ceremonially* given to members of own clan.

(*a*) When a woman dies, her brother will go to the corpse and will dress her with a skirt, paint her face, and attend to the details of her funeral. He will then say, "I am the man who will call the names."

(*b*) When a man dies, his son may go to the dead man's sister in her house shortly after the death. He puts a shell valuable, generally a *Turbo* shell, into water in one of her pots and he cuts her hair for her. The shell becomes her property and the brother's son gains the right to give the names to his children. (In the actual case in which this procedure was followed, the woman had no other surviving brothers.)

(*c*) When a man dies, the widow's hair will be cut and a shell valuable presented to her in water by either her brother or her brother's son. (In the texts in which this procedure is mentioned there is, however, no mention of the right to the names and it is possible that in this context the ceremonial is merely an affirmation of the widow's membership of the clan into which she was born but from which she has been to some extent separated by marriage.)

(*d*) If a woman is left the solitary survivor of her clan, all the names of that clan become vested in her and her bride price becomes correspondingly great since the right to give the names will fall to her husband or her children. In such cases, an effort is made by the members of closely related clans, her distant classificatory brothers, to get her as wife for one of them, so as to retain the names within the phratry. In this procedure we may see weak identification, based only upon classificatory brotherhood, being strengthened by the additional identification based upon marriage.[1]

---

[1] As might be expected from the complexity of these observances, quarrels sometimes arise over the matter of names held by women. I came across one dispute of this sort: The two last male members of the Mbe clan had been killed by a raiding party some years before, leaving only two women. An unrelated man, Tshamelwan, had provided the food (? or valuables) for the mortuary feast for these men and on the strength of this he had claimed—"I take the names!" and had proceeded to give Mbe names to his sons.

But there still remained the two women, by name Teli and Tampiam. Teli was an old widow, with two sons to whom she had given some Mbe names. Tampiam was a young woman.

The clan, Iavo, closely linked with Mbe, therefore arranged that

5. That a woman is to some extent identified with her husband. In this case we have to consider an identification which is not reciprocal. It seems that the wife does take over a part of her husband's status, but it seems that he takes over little, if any, of hers: an asymmetry which is consistent with preponderant patriliny and patrilocal marriage. In our analysis of the *naven* ceremonial we shall, however, be concerned only with cases of wives taking status from their husbands; and that only *à propos* of *mbora* (mother's brother's wife) and *tshaishi* (elder brother's wife). This formulation is not relevant to the patterning of *wau*'s behaviour in *naven*.

The identification of wife with husband is in general very little summarised in the culture. We do not find, for example, any tricks of the naming system in which this identification is expressed or assumed. Neither partner changes his or her name at marriage; and, though new kinship terms may be applied to the partners as a result of the marriage, these new terms for the most part still discriminate between the partners. This is true of all the marriages of closer relatives, including all marriages of siblings of the parents and of persons who are identified with these in classificatory terminology. The sole exception to this among the near relatives is the use of the term *naisagut* for wife's father and his wives.

But in the less important ramifications of kinship we find that the wife is referred to by the same term as the husband, e.g. *mbuambo* is the term used for mother's father and his wives, as well as for mother's brother's son and his wives. Similarly *kaishe-ndo* and *kaishe-ragwa*, terms differing only in gender termination, are used reciprocally between the parents of a husband and the parents of his wife.

In the case of *tshaishi* (elder brother's wife) the terminology is interesting. She always identifies herself with her husband

Tampiam should marry Tshava (my informant), a young man of Iavo, so as to keep the Mbe names in the phratry.

No children had been born of this marriage when I was in Kankanamun and Tampiam was still angry with Tshamelwan for stealing the Mbe names. Lacking children, she was doing her best to reserve the names by giving them to her pigs, canoe, etc. She also attributed her sterility to sorcery and suspected Tshamelwan of being the cause of it.

Tshava stated that Tshamelwan repudiated this charge, suggesting that the sterility was caused by Teli.

in calling his younger brother *tshuambo*, a term which is otherwise reserved for younger siblings of the same sex as the speaker. The ordinary reciprocal to *tshuambo* is *nyamun* (elder sibling of same sex), but the husband's younger brother does not use this term for his *tshaishi, except in the* naven *ceremonial.*

The identification of man and wife is somewhat more evident in everyday life than in the terminology of kinship. As we have seen, husband and wife are discriminated in their contacts with near kin but identified when they are dealing with outsiders. The picture of everyday life fits with this, inasmuch as the contexts in which we see the couple identified are especially those in which they are dealing with outside persons or inanimate objects. We find them, for example, working together, sharing a house[1] and acting as each other's agents in barter. There is a very close mutual economic dependence between man and wife; and though every piece of property can be defined as belonging either to him or to her, it is almost possible to see the household as a single economic unit. This may be illustrated by the case of a man in Komindimbit village who had lost all his own and his wife's property in a fire which had destroyed his house. I asked him how he would reinstate himself and he replied immediately, "My wife will feed up pigs and I shall sell them". Similarly, if a wife comes without dowry to the household, her co-wives will hold it against her, as a reproach, that she contributed nothing to the common wealth of the group.[2]

[1] It may be noted that the husband is often absent from the house and spends much of his time in the ceremonial house with the other men. It is not approved of in a man that he should spend a great part of the day in his dwelling house among the womenfolk.

[2] At this point we may note a weakness in the argument. We shall later regard the behaviour of *mbora* and *tshaishi* as consistent with this formulation. But I am not able to document the identification of man and wife in those contexts in which they are dealing with the husband's younger brother and the husband's sister's son. Man and wife *are* identified when they are dealing with outside people but it is not clear that they are identified in those contexts which are specially relevant to the present enquiry. The concept of *context* must certainly be added to that of *identification* and we should be chary of using the latter term without some indication of the contexts in which the persons are identified.

CHAPTER V

Sorcery and Vengeance

IN the previous chapter I have stated that in the Iatmul kinship system there are four identifications which we may regard as major premises of the system:

1. An identification between father and son.
2. An identification (on different lines) between a boy and his mother's clan.
3. An identification between brother and sister.
4. An identification between husband and wife.

The working of these premises will be illustrated in the present chapter from material which I collected about a special context of Iatmul culture, that of retribution.

In the investigation of Iatmul ideas about sorcery and retribution, we at once come upon the native concept, *ngglambi*, a word which we might translate into abstract terms as "dangerous and infectious guilt". But in native thought *ngglambi* is much more concrete than this. It is thought of as a dark cloud which envelops a man's house when he has committed some outrage. This cloud can be seen by certain specialists who, when they are consulted as to the cause of some illness or disaster, rub their eyes with the white undersides of the leaves of a tree, and are then able to see the dark cloud hovering over the house of the person whose guilt is responsible for the sickness. Other specialists are able to smell *ngglambi* and say that it has a "smell of death—like a dead snake". It is characteristic of *ngglambi* that it may "turn and go", causing the sickness either of the guilty person or of his relatives. It is this transmission of *ngglambi* which is relevant to the questions of identification between kin.

As to the agency responsible for the dangers of *ngglambi*, the people are very articulate and their statements fall into four closely related groups.[1]

[1] There is a fifth phrasing of the matter which, though irrelevant to the validity of *ngglambi* infection as an index of identification, is yet interesting for the light it throws upon Iatmul sorcery. It is believed

(*a*) It is said that the *ngglambi* itself causes illness and death. This is an absolutely impersonal theory of the agentless causation of disease and is tantamount to the statement, which the Iatmul also make: that a man's sorcery will of itself return to plague him. This type of phrasing may almost be regarded as a euphemism for the more personal statements which follow.

(*b*) It is said that the illness is caused by certain spirits, *wagan*, who avenge various offences, acting upon their own volition but generally in the interests of their clan descendants.

(*c*) It is said that the persons whom the guilty man has offended will complain to their clan *wagan*, who will avenge the offence. Such complaints may take various forms: in some cases it is apparently sufficient if the *wagan* hears the weeping of his descendants; when he hears he will at once set about avenging the injury which has been done to them. Failing this, the offended people will beat a rhythm on the *nggambut* (mortuary post) of the person whom they wish avenged. This rhythm would be the slit-gong call (*tawet*) for summoning the *wagan*. The latter would hear the rhythm "and having heard it he would not forget. There would be no talk about the matter. To-morrow the *wagan* would come to smell out the *ngglambi*". Alternatively the offended people might go to the *wagan* personally at a time when he possesses his shaman. They would talk to the possessed man (i.e. to the *wagan*) and pay him areca nuts and shell valuables for his services in avenging the offence.

(*d*) It is said that the offended persons will themselves practise sorcery which will cause the sickness or death of the guilty person or of his relative. Alternatively they may pay a professional sorcerer to do this with or without the aid of his *wagan*.[1]

_____

that a cunning sorcerer will kill off victims *in pairs*. One of the killed will be the person against whom the sorcerer has a grudge, while for the other he will choose one of his own relatives—thus "short-circuiting" the retribution and reserving to himself some choice as to which of his relatives shall die to pay for the desired death of his enemy.

[1] In this community where violent death is frequent, death by sickness (i.e. sorcery) is not very seriously regarded. There is very little of the almost paranoid fear of sorcery which is characteristic of many primitive cultures.

In all these cases the resulting sickness may be described as due to the *ngglambi* of the guilty man. Thus the theory of *ngglambi* fluctuates between a belief in the spontaneous effects of guilt which may be "infectious" and concrete statements that so-and-so will avenge himself either upon the offender or upon his relatives.

In the light of this double meaning of *ngglambi* it is interesting to see which of a guilty man's relatives are most likely to be imperilled by this dark cloud—since we may reasonably translate a statement, that *A*'s *ngglambi* has caused the sickness of *B*, into our jargon of identification: "*A* has offended against *C*; *C* identifies *B* with *A* and proceeds to avenge himself upon *A* by practising sorcery upon *B*."

But though this would seem to be the natural interpretation of *ngglambi* infection, the theories of retribution contain another emphasis which would, at first sight, seem to confuse the issue. This is the emphasis upon the law of "an eye for an eye and a tooth for a tooth". For example, if *A* magically kills *B*'s wife, *B* will be angry and will avenge himself by killing *A*'s wife. Such an exact retribution the Iatmul would approve of;[1] but they would still phrase the death of *A*'s wife as due to *A*'s *ngglambi*, although in killing the woman *B* was not identifying her with her husband. But such a confusion of the issue is, I think, more apparent than real. The death of *A*'s wife is still due to an identification of a husband with his wife: *B* identified himself with his wife sufficiently to avenge her death, and the *lex talionis* indicated that the correct person for him to kill was *A*'s wife. In place of the

---

[1] In such a case, the avenger will first perform the magic in secret, but when the woman is dead he will stand up in the ceremonial house and proclaim what he has done and his justification for doing it. If the *lex talionis* has been accurately followed, this may be the end of the matter; otherwise a long sorcery feud may result accounting for as many as fifteen deaths.

As regards the effectiveness of sorcery and *ngglambi* in causing sickness and death, I have adopted the native phrasing of the matter. From a European point of view it is probable that most of the deaths in a Iatmul sorcery feud are due to "natural causes" and the feud itself may be almost fictitious. But in native belief, the deaths are ascribed to the progress of the feud; and the fact that the deaths do not usually follow the *lex talionis* with any accuracy is ascribed to the unreasonableness of *wagan* and sorcerers.

question "Who is likely to be identified with the guilty man?" we have only substituted the question "Who is likely to identify himself with the killed person?" The phrasing of retribution in terms of *lex talionis* when analysed in this way justifies us in regarding the *ngglambi* infection which occurs between two relatives as an indication that persons related in the same manner are to some extent identified.

An additional piece of evidence which seems to corroborate the theory that *ngglambi* and *lex talionis* may be equated from a structural point of view is afforded by the homonymous meanings of the word *nggambwa*. This word means on the one hand "vengeance", in such phrases as *nggambwa kela*, "to take vengeance"—the duty of the man whose relative has been killed by enemies. But on the other hand the same word is also applied to certain spirits, *kerega nggambwa*, literally "eating vengeances", who after death tear the flesh from the corpse with their long finger-nails. Thus in the word *nggambwa*, as in the word *ngglambi*, we find a supernatural concept mixed up with concrete ideas of vengeance.

With this preliminary examination of the native ideas about sorcery and vengeance, we may now proceed to examine a series of stories of sickness and death, confident that when we find one individual suffering for the deeds of another we are entitled to see in that fact an instance of identification either between the guilty person and the sufferer or between two persons related in the same way as these.

The material here presented is, strictly speaking, only relevant in its present context for the light which it throws upon the various identifications, but inasmuch as the stories are also interesting from other points of view I shall give them in rather more detail than would be necessary merely to document the identifications.

Case 1 is a translation of a dictated text; but the other stories are reconstructed out of very full notes taken during slow narration of the stories in pidgin English or in the native language or in a mixture of the two. I have not scrupled to insert conjunctions and binding clauses in these texts so as to make them more intelligible, but wherever a more important explanatory statement has been inserted, it is put in parentheses. Names of the people concerned are given in somewhat shortened form.

Case 1. This is an extract from a general account of the proceedings which would accompany the sickness and death of a man. The account was given me by Mbana in Kankanamun. It describes the procedure of divination for the cause of the sickness:

His (the sick man's) son and wife take ashes from the fire basin and set them in little heaps on the spathe of a palm. There will be several heaps on the one spathe (one heap for each relative whose offences may be the cause of the sickness). Then they say, "What is the root of the matter? Why did he die? Was it because his father went to other people's wives? Was it because of the doings of his father's father that he died?"

Then the palm spathe with the heaps of ashes is put into a mosquito bag. The *wagan* will press the ashes with his hand. Then they take the spathe out of the mosquito bag (and look to see which heap has been disturbed) and they say, "Yes, it is his father's fault."

This extract would seem to indicate that *ngglambi* infection comes typically down the patrilineal line and illustrates the identification of father and son.

Case 2. In a conversation with Tshava of Kankanamun I enquired whether *lando* (sister's husband) could be affected by *tawontu*'s (wife's brother's) *ngglambi*. My informant told me that such infection would always pass to the *lando*[1] if he gave food to his wife's brother, and in general that befriending a sorcerer was dangerous because the sorcerer's *ngglambi* was liable to affect those who gave him food or lodging. He said that if any man saw another eating a particular creeper used in sorcery, he would spread a warning: "Do not give food to that man." But the danger is only attached to gifts of food and not to gifts of valuables.

---

[1] It appeared from Tshava's remarks that though *ngglambi* from *tawontu* might infect *lando*, the reverse was not the case. But this emphasis upon the one direction may have been due to the experiences of his family in its relationship with Malikindjin.

"If you give food to a sorcerer your pig may die, or your dog may die, or your child may die."

In illustration of this, my informant quoted the case of his parents: his father Djuai had married Malikindjin's[1] own sister, Nyakala. She was always giving food to her brother and Djuai often scolded her for this. Because she befriended the sorcerer "many" of her children died, and finally she herself died. Later, Nauyambun, a very old woman, the mother of Malikindjin, came to the house and my informant was constantly sick (because of this connection with the sorcerer).

Returning to the question of his mother's death, Tshava narrated to me in some detail how he and Malikindjin had avenged her:[2]

When Malikindjin was mourning for Nyakala he came to me and said to me, "We will go down to such and such a piece of bush." We went and there found a man named Tamwia. Malikindjin said, "Now we are going to take vengeance for your mother on this man." He told me to hide myself and he used magic to make Tamwia unable to see him. Tamwia was "in the dark" and Malikindjin made him "cold". He did not cry out. Malikindjin came straight up to him and broke his neck and head with an adze. He killed him. He cut off his head and put in its place a head made of *nggelakavwi* (a tuberous plant, *Myrmecodia* sp.). He threw the real head away.

We each took an arm and hid the body in the elephant grass. Malikindjin said, "Don't talk about this, but there

---

[1] For a description of the personality of the old sorcerer, Malikindjin, cf. p. 164.

[2] The magical incident which is here described is very closely comparable with one collected by Dr Fortune in Dobu. In both incidents the informant acted as watch-dog and the magic was of the type called "vada". In it the victim is first killed and then brought back to life only to die later. The two stories differ in that in Dobu it is the heart of the victim which is removed instead of the head, but I was told on another occasion that the Iatmul sorcerers sometimes remove the heart. And they differ too in this, that Dr Fortune's informant was moved almost to hysteria in the narration of the story, whereas my informant described the events in as cool and as detached a manner as he would have described a magical procedure for the improvement of yams. (Cf. *Sorcerers of Dobu*, 1932, pp. 158–164 and 284–287.)

is no fear of trouble. We are only taking vengeance." Then I went away to work on Malikindjin's garden, and came back to find the man Tamwia lying sick and groaning. Malikindjin woke him up and said, "Are you asleep?" and Tamwia said, "Yes." And Malikindjin said, "Can you see me?" Tamwia said, "Yes." Malikindjin said, "What have I been doing to you?" and Tamwia said, "I was asleep and you woke me up." Malikindjin said, "Right. That is the way to talk. Now go away."

Malikindjin also told him, "You cannot cry out with sickness soon. You will stay awhile first", and he gave him a date—five days thence.[1] "You cannot mention me. It is not good for you to live in Kankanamun. You had better die." Malikindjin addressed Tamwia by his -*awan* name.[2]

Then we went back to the village and Malikindjin said to Djuai, "Do not wear mourning all the time. I have taken vengeance. You can take off your mourning, or only wear a little." The poor man, Tamwia, soon died.

After this Malikindjin wanted to teach me his spells but I was afraid. If I paid him with valuables and they were not enough, he might kill me. Later, Malikindjin debated the matter in the ceremonial house. He said, "My sister is balanced by a man. Why did you send your *wagan* to kill my sister? But now we are avenged."

But Kwongu sent his *tshumbuk*[3] (personified pointing stick) to kill one of Malikindjin's wives and she died, and Malikindjin killed Kwongu's wife and said, "You killed my wife, now I avenge it. There is no reason why I should hide the matter."

Then Kwongu killed a child of Malikindjin and Malikindjin killed the wife of Iaremei. Iaremei complained of this and said, "Why do you kill my wife?" but Malikindjin said, "Kwongu killed my child and you have given Kwongu food. Why do you give him food? Now his trouble has come to

---

[1] Among the Iatmul, in magical transactions when a date is fixed it is invariably *five* days hence.

[2] I had to ask a leading question to get this detail. But I was on other occasions told spontaneously that the -*awan* name is used in addressing the victim of sorcery, or his image (cf. footnote, p. 43).

[3] A *tshumbuk* is a pointing stick, generally a shortened spear shaft. It is personified and the only specimen which I saw had a face worked in clay in the middle of the shaft.

you and your wife has died." Iaremei said, "True, I was stupid." He did not avenge his wife's death but went away to Timbunke village, where later he was killed in war by natives of Kararau.

From this story we may extract the following identifications, which we may phrase in terms of *ngglambi*:

(*a*) The passage of *ngglambi* from the sorcerer to his sister's husband; i.e. identification of a man with his sister and of the sister with her husband.

(*b*) The passage of *ngglambi* from the sorcerer to his mother and thence to those who befriended her.

(*c*) The passage of *ngglambi* from husband to wife (*A* kills *B*'s wife and *B* kills *A*'s wife).

(*d*) The passage of *ngglambi* from father to son.

(*e*) The passage of *ngglambi* from the man who befriends the sorcerer to his wife.

Case 3. My informant Tshava of Kankanamun gave me another instance of the dangers of befriending the sorcerer. He said that Tshe always gives food to Malikindjin and that is why his child is always sick and many pigs have been killed.

Once Malikindjin said, "I have had a dream. Somebody will have to kill a pig" (i.e. he had dreamt of *ngglambi* or some disaster and a pig would have to be killed to propitiate the *wagan*). Then there was a debate against Malikindjin and a number of men said, "This is to be Tshe's pig, he is going to pay now." And they said to Tshe, "You are always giving food to Malikindjin and now you will get his trouble." And Djuai (Malikindjin's sister's husband) said, "Always Malikindjin's sister caused trouble in my house, and now Malikindjin has gone to your house and your children will die. If anyone kills a pig you will have to pay for it" (i.e. there was a danger that if to avoid the disaster somebody had to kill a pig, Tshe would be held partly to blame; and therefore unless he paid for the pig his children would die). But Tshe said, "Malikindjin is my father (classificatory); I cannot not help him."[1]

[1] The avoidance between father and son, whereby the father must never eat food produced by the son, is only observed by *own* father and

In this instance, we have a very clear illustration of the passage of *ngglambi* to a man who befriends the sorcerer because he regards him as a "father". The man is even expected to pay in advance for the effects of the sorcerer's evil deeds in order to ward off the danger which will otherwise descend upon his own children.

Case 4. This is a case of identification of husband and wife of which I heard when collecting census material in Palimbai.

The husband had assisted the Australian government in putting down head-hunting. He and another man had used their influence to make the natives of Palimbai burn their spears. Later, his wife died and her belly was much swollen. They found a piece of a burnt spear in it. (Probably this fragment was produced in a conjuring trick performed by a medical practitioner.) This was regarded as proof that she had died from sorcery resulting from the burning of the spears.

Case 5. Informant: Gimelwan of Palimbai. A quarrel arose between Djuai[1] and Pandu over the tidying of the ceremonial house. Djuai was sweeping the ceremonial house and picking up the rubbish with a pair of planks. He was cross with Pandu because he was doing nothing. Pandu kicked Djuai and Djuai then killed Pandu "with a small stick and broke his belly". Pandu was buried.

Lapndava, the son of Pandu, threw a spear into Djuai's house (a gesture of anger and challenge). Djuai then presented twenty shell valuables of various sorts to Lapndava.

But the matter did not finish there. Later, Djuai's "children died and his wife and now he is all alone" (the implication being that the wife and children died from sorcery following upon the quarrel).

On the other side Lenagwan, Pandu's elder brother, was pursued by misfortune. "He curled his lip in Kankanamun" (i.e. he sulked). He wanted to marry a woman in Kan-

---

father's *tambinyen*. It is correct for a man to give food to a distant classificatory father (in this case, father's mother's sister's son). Moreover the taboo is relaxed when the father is very old.

[1] The Djuai of this story was a man of Palimbai, not the father of Tshava who has the same name and is mentioned in the other stories.

kanamun but could not. Then he went to the debating stool in the Kankanamun ceremonial house, and called the men out to raid Kararau, and they went out to shoot the women while they were fishing on the Kararau lake. Lenagwan threw his spears but killed no one and the raid was a failure. Later, the people of Kararau organised another raid in reply and killed three women of Kankanamun, and finally Lenagwan was killed by sorcery done by the men of Kanakanamun because he had organised the ineffectual raid. "Therefore Pandu's clan is now almost finished." And my informant mentioned two other members of the clan who had died, one of them killed by Europeans and the other killed by sorcery.

This case contains identifications of father and son, man and wife, elder brother and younger brother, and a vague reference to identification based upon clan membership at the end of the story.

Case 6. Informant: Waindjamali of Palimbai. This story was given me in explanation of a shamanic séance which I had recently witnessed. The séance was held to enquire into the death of Tepmanagwan, the great man whose mortuary image is described later (p. 154). The death was the latest result of a feud which had started between rival initiatory grades some four or five years earlier. My informant was a member of grade *By* 4 (cf. Diagram, p. 245); Tepmanagwan was a member of *By* 2 and the quarrel, which started between *Ax* 3 and *By* 4, spread to include almost the whole of *Ax* and *By*.

The *bandi* (*Ax* 3) called us (*kamberail*, *By* 4) out to fight. They beat on the slit-gongs and insulted our fathers. "You children are trash and bastards. We have played with your mothers." Then there followed a brawl with no spears but only sticks. We were beaten, but we said, "We will avenge it on their children" (*Ax* 5). Then the other moiety made us buy pigs and we presented them. Kumbwi (a man of *Ax* 1) and his brother Malishui were conspicuous in this brawl. Kumbwi beat a slit-gong[1] to call his moiety together

---

[1] In descriptions of brawls, it is usually stated who beat the gongs; and from Case 7 it would appear that the man who beats the gongs is considered responsible for the brawl.

and carried a spear. Malishui only carried a stick. Then Tepmanagwan came up and contradicted Kumbwi. Tepmanagwan picked up a spear with barbs (such as nominally should not be used in brawls inside the village) and he speared Malishui in the leg. Then all the opposite moiety combined against Tepmanagwan, but his own moiety came to his help and he was also helped by his classificatory *laua* and his own daughter's husband. After the brawl Tepmanagwan said, "So far they have always bullied us; tomorrow we shall fight. So far we have always used bamboo points; to-morrow we shall use barbs." Next day there was a brawl and Tepmanagwan was wounded "with many spears" and that night we all scattered. I and others of the Kamberail grade (*By* 4) went to Tegowi village. Tepmanagwan went to Malikindjin, his classificatory brother in Kankanamun. He gave a mother-of-pearl crescent and conus disks to the old sorcerer—and Malishui died.

Soon after that Wandem (Tepmanagwan's younger brother's wife) died. "Malikindjin killed her with a *tshumbuk* (personified pointing stick) as retribution for Malishui. He killed her 'in return' because he was afraid they would turn the *ngglambi* back on to himself or his relatives."

Then Kumbwi wept and gave valuables to his *wagan* (ancestral spirit) and said, "Ancestor, my younger brother only has died, but they shall not stay unhurt."

Then two sons of Tepmanagwan died, one of them while he was away as an indentured labourer.

Then there followed five other deaths.[1] I tried to get information as to the exact relevance of these deaths to the feud, but my informant dismissed the matter, "They are not straight. They only turn about. That is Malikindjin's way."

Finally Kumbwi gave a mother-of-pearl present to his *wagan* and Tepmanagwan died; and in the séance enquiring into his death the *wagan* (i.e. the shaman in a state of

---

[1] Probably most of the deaths in this sorcery feud were due to influenza. But we are here concerned, not with the theories of European medicine, but with those of the Iatmul. In general *no* deaths are regarded by the latter as "natural" but all are referred to sorcery. Even when a man is killed by the enemy, his relatives will suspect that some sorcerer in the village has sold the dead man's soul to the enemy before the raid.

possession) produced this mother-of-pearl crescent and said, "This is the shell that killed Tepmanagwan", and he smashed it against the post of the ceremonial house as a sign that that feud was finished.

This case contains first of all an example of solidarity within the initiatory moieties, but the ties between *laua* and *wau* and between son-in-law and father-in-law are given preference over allegiance to the moiety. In the sorcery feud we have identifications of brother and brother, of younger brother's wife and husband's elder brother, and of sons and father.

Case 7. This is the story of a long sorcery feud in Kankanamun which was given me by my informant, Tshava. As in the case of the similar feud in Palimbai, this series of deaths had its beginning in a quarrel between initiatory moieties:

A boy was playing the Pan-pipes in the junior ceremonial house and a woman, Ialegwesh, came up outside the house. The boy said, "Come up. I want to see your sore."[1] She went away but later came back to the junior ceremonial house and shouted, "My *windjimbu*,[2] I want to see you." Then the boy hid and went to his house. Later Ialegwesh was heard playing the Pan-pipe tunes on a jews' harp, and the boy came with others and scolded her. She said, "What, do you think I don't know the Pan-pipe tunes? Mwaimali (her husband) taught me them." A senior man, Katka, overheard this and was angry. He went off and told the other men and chased Ialegwesh with a stick. He found some men beating on a suspended log.[3] He said, "Stop doing that. A woman has exposed all our secrets." Then Djindjime beat a summons on the slit-gong and all the men came together. They went and raided Ialegwesh's house and broke her pots; but her brothers came to her assistance and there was a brawl between them and Katka. At first they were many and he was only

[1] An obscene metaphor for female genitals.

[2] The various secret sound-producing objects, flutes, Pan-pipes, etc. are regarded as connected with the wood spirits or *windjimbu*. There are many stories of women who have had love affairs with *windjimbu*.

[3] This is a junior analogue of the *wagan* slit-gongs, an initiatory secret. The tunes for the Pan-pipes are also secret.

helped by the small boys (of the junior ceremonial house). Katka fell in the brawl and then others came to his assistance. His eye was broken.

Next day Malikindjin came and saw Katka's eye and asked what had happened. Katka said, "We had a brawl." Malikindjin said, "Why?" and Katka said, "About men's things" (i.e. initiatory secrets). Then they debated the matter and Malikindjin took the side of Ialegwesh and Mwaimali. He said, "Somebody will die from this trouble." Later, he sent his *tshumbuk* (personified pointing stick) and killed Kwongkun.

On the day after this death, some young men came home from indentured labour with their boxes full of trade goods. Malikindjin went to the ceremonial house and demanded a box of trade goods from the labourers. "I have been looking after your women while you were away. You must pay me one box." The boys paid him, but they said, "That box will eat up a man later."

The box ate Katka (i.e. Katka died of sorcery because it was against him that Ialegwesh had needed Malikindjin's help). Katka's death was avenged on Ioli and his younger brother, Kwilegi, who had helped Ialegwesh. These two died in one night (this was probably during an influenza epidemic). Later, Katka's wife died to balance Kwilegi. Then the *wagan* said, "Who else helped Katka?" and Djindjimowe died, and to balance him Kisak.

Then the *wagan* said, "Who beat the slit-gongs?" and Djindjime died. Finally Ialegwesh died, killed by Malikindjin's spells. The old sorcerer was acting on both sides. The beginning of the trouble lay with her and that is why she died, and they all said, "That talk is finished. It is done with."

But later Mwaimali, Ialegwesh's husband, married Kagan, the widow of Katka, and she bore him a child. Then Malikindjin's *wagan*, Tshugu, said, "Yes, at first a woman made trouble and Katka died. Now you have married Katka's wife." Then the child died.

Then Mwaimbwan, a member of Mwaimali's clan, said, "This affair was done with long ago, but now you want to start it again" and he caused a child to die on the other side.

Finally there was a séance. Malikindjin went into a trance and Tshugu (speaking through Malikindjin's mouth) called upon all who had been concerned in this feud to kill a big pig to finish the feud. He asked for the names of all who had taken part in the original brawl on each side, and told the members of both sides to contribute valuables. A pig was bought and killed and the various *wagan* (in their shamans) and all the men ate it. After the feast all the *wagans* took the bones of the pig and they made a hole close to the ladder of the ceremonial house and there buried them, at the suggestion of Tshugu; and they said, "If any man raises this feud again, he shall die."

But later Malikindjin fell sick.[1] That is because everybody wants vengeance upon him. All who know anything about spells and all the *wagan*, they are all agreed.

In this story the following identifications are recognisable:

(*a*) Identifications based either on sex solidarity or on membership of initiatory moieties (cf. the case quoted on p. 99, in which a woman was detected spying upon the flutes).

(*b*) A woman's brothers side with her, giving precedence to that allegiance over their identification with their sex or moiety.

(*c*) Those who help the offending woman suffer, including a pair of brothers.

(*d*) Katka's wife dies from her husband's *ngglambi*.

(*e*) Finally the marriage of Mwaimali to Katka's widow provides such a tangled position that it is not clear whether the child that was born to this marriage died because of Ialegwesh's sins or those of Katka.

Case 8. Tshava of Kankanamun told me the following story about the death of two of his father's wives, Mait and Undemai:

Two of Malikindjin's wives, Mwaim and Lemboin, were quarrelling. Mwaim said, "Djuai (my informant's father) is

---

[1] He died about four months later. I enquired about the burying of the bones of the pig and was told that this was not a customary procedure, but was done simply at the *wagan*'s suggestion.

always playing[1] with you." Malikindjin heard of this and
fetched help from Tegowi village to avenge himself on
Djuai; and there was a brawl in Kankanamun. Later, Djuai
killed a big pig and gave it to the men of Palimbai, inviting
them to come and kill Iowimet, a native of Tegowi who was
settled in Kankanamun (Iowimet and Malikindjin were classificatory brothers). But Iowimet fled to Tegowi and they
only destroyed his property and stole his shell valuables.
Later he came back and wept for his things. He had a string
bag, a melo shell, a *Conus* shell necklace and a *Turbo* shell.
These he gave to Malikindjin. Malikindjin said, "Why do
you give them to me?" Iowimet said, "I am crying for my
things. They have killed my pig, etc. I want one of their
women to die." Then the two of them agreed on a day—"on
this day we will meet in the bush".

On the date, they met and they waited for Mait (Djuai's
wife) at the place called Tungwimali on the river bank. When
she came they lied to her and said, "There are crocodiles'
eggs here", and when she came ashore they killed her—
killed her completely and revived her. She went back and
told my father (Djuai), "It was those two who killed me.
They lied about crocodiles' eggs and I went ashore." Djuai
only heard what she said and then she died. Djuai wept for
her and called upon his *wagan*, Tshugu.[2] He said, "I am the
only one who is weeping. They are untouched. Later they
must cry too." Then Tshugu killed Mwaim, the wife of
Malikindjin.

Later there was a séance and Tshugu possessed Malikindjin
and spoke addressing Malikindjin, "You, my shaman, why
are you mad and kill all the women? Now they are taking
vengeance. My clansman called on me and I came and killed
the woman. One of your wives has died. Later somebody
belonging to Iowimet shall die." Then Yivet, Iowimet's
daughter's husband, died. Then Malikindjin and Iowimet
started again and killed Undemai, another wife of my father.
Then it was all balanced.

[1] A euphemism for copulation.
[2] This was the *wagan* which possessed Malikindjin. My informant
explained that when Djuai called upon Tshugu it was only the spirit
who heard, not the shaman.

But Iowimet did spells on an adze and found one of my footprints. He hit the ground with the adze and then wrapped up the adze in nettles and put it in a pot and made it hot on the fire. My sore hurt me and I cried out. Then Iowimet put a little water on the adze and cooled it, and my sore was cooled. This went on till I took a *Turbo* shell and my wife carried it to Iowimet. Then Iowimet said, "You have paid me and later the sore will dry up." It dried a little, but the doctor (i.e. the Government medical official) took me up to Ambunti (to the Government hospital). The doctor said, "Why is the sore so big?" And I said, "It is only Iowimet." The doctor said, "What was the quarrel about?" And I said, "It was my father's quarrel, but it is not right for the son to help his father."[1] Then the doctor sent a police boy to get Iowimet and asked him, "Why did you hurt this boy?" Iowimet said he did not know, and the doctor said, "Tshava has told me. Why are you lying?"

Iowimet was nearly put in prison, but he is my (classificatory) *wau*. The doctor asked me if I was sorry for him and I said "Yes", and the doctor said "Never mind" and dismissed the matter.

In this story the following identifications occur:

(*a*) Malikindjin and Iowimet; brother with classificatory brother.

(*b*) Malikindjin and Mwaim; husband and wife.

(*c*) Djuai and Mait; husband and wife.

(*d*) Iowimet and his daughter's husband.

(*e*) Djuai and Undemai; husband and wife.

(*f*) Djuai and Tshava; father and son.

(*g*) And finally Tshava "is sorry for" his classificatory *wau*.

Case 9. This story was given me by Tshava of Kankanamun. It describes the vengeance of the *wagan* upon a medical practitioner who interfered in their affairs and demanded no payment for his services:

Mongwaevi, the *wagan* of Yivom's clan, had told him to kill a fowl, but Yivom said, "Why should I? When should

---

[1] Cf. the taboo upon the father's eating any food grown or prepared by the son.

I?" and refused. Then the *wagan* turned and made Yivom's small son sick, and Tengai, a man with some knowledge of spells, cured him. But the *wagan* said, "It was not his (Tengai's) affair. It was a case of which people had said 'he is going to die', but that man Tengai used spells."

Then the *wagan* made Tengai's wife sick. The people said, "Yes, they paid Tengai nothing for his spells and he did not ask for pay (i.e. inasmuch as he had not been paid it was not his business to interfere)." Punpunbi's daughter said, "It is witches" and "Yivom's folk will not stay untouched. Our *wagan* will look at them. It is their *ngglambi*." Punpunbi agreed with her daughter's opinion and she summoned her husband Tengai and was cross with him. She sent him away from her; though before that they had agreed well enough together. For a long while they did not summon the *wagan*, but finally they called for Mongwaevi's shaman who went into a state of possession and the *wagan* demanded areca nuts from Tengai. Only then, after the presentation of areca nuts, did Mongwaevi clear the road to the land of the dead. The soul had wanted to go before but Mongwaevi (in spite) was closing the road. The woman was terribly sick but could not die. Then she died.

In this case Yivom's original offence against Mongwaevi is first avenged upon his child (identification of father and son); then the *ngglambi* passes to Tengai—an example of the danger of helping those with *ngglambi*. Then from Tengai the *ngglambi* passes to his wife.

Case 10. This story was related to me by Tshava of Kankanamun. I have rearranged the story so that the sequence of events may be clear:

Atndjin married a woman named Tshugwa, and later she helped him to marry her younger sister, Wama, by contributing to the bride price valuables which she had got by breeding pigs. My informant told me that it is quite usual for an elder sister to help her husband to marry her younger sister.

Later Atndjin died, leaving the two sisters as widows. According to custom they should have become wives of

Atndjin's younger brother, but this did not occur. A man named Tshuat of Jentschan wanted to marry the elder sister and finally took her to Jentschan, but Malikindjin wanted her and sent his *tshumbuk* (personified pointing stick). Tshuat saw the *tshumbuk* in a dream and was frightened, so he let her go. Malikindjin took her and married her and paid a bride price to Atndjin's younger brother. Then Malikindjin said, "When the younger sister marries I shall get part of the bride price—as much as the elder sister contributed— never mind about a pig—or if there are not enough valuables, let them pay a pig."

But the various husbands of the younger sister endeavoured to dodge this payment. First a man named Tevwa took her, but he became sick and was frightened of Malikindjin and sent her away. For a while she remained unattached till Tshaunanti took her. But his other wife died because of her; Malikindjin sent his *tshumbuk* because the pay was inadequate. They said, "Tshugwa's things (i.e. valuables) have eaten up your wife." Then Tshaunanti sent her away. Then Mbulm- buangga married her, but his child by a previous wife died because he did not kill a pig for Malikindjin. Though his child had died he still did not pay the sorcerer. Malikindjin was an old man and Mbulmbuangga said, "He is no good any more. He used to be a hot man, but he is no good any more."

In this case we see first a husband, Malikindjin, identifying himself with his wife and on that ground demanding recom- pense for valuables which she had expended. Later, we see the offences of successive husbands of the younger sister being avenged (1) on the husband himself, (2) on the hus- band's other wife, and (3) on the husband's child by a previous wife.

Case 11. This is a story which was told me by Tshava of Kankanamun. The story is irrelevant to the question of identifications with which we are immediately concerned, but it is of interest, after this series of feuds, to cite a case in which a man killed his wife and afterwards received homicidal ornaments from her relatives:

Laindjin killed Tualesh, his wife. She was a member of Wolgem clan and (therefore) my (classificatory) sister.

Tualesh went to get water in a long bamboo. When she returned Laindjin said, "Why were you so long?" She said, "I was not long. What are you thinking?" Laindjin said, "I was only asking," and she said, "Other men don't ask, but you do. You are jealous." Laindjin said, "Yes, I know how women carry on." Then he jumped up and seized the water bamboo and bashed her with it and she died. They made incisions in her skin with a bamboo knife and used spells, but they could not cure her. Then Laindjin wept.

The members of Wolgem clan heard of it and we said, "Our sister is dead." Then we took spears and went to Laindjin's house. He had blocked the entrance (probably with a mat) and was inside. Then we flung many spears at the house and Laindjin's brother's son was wounded, but not killed.

We presented *tambointsha* (homicidal tassels), black paint and Crinum leaves to Laindjin and he killed a pig for the *wagan* and the other men, but we did not eat it; it was our sister's pig.

Later, Laindjin said, "This woman did not die without a cause. Malikindjin killed her." And he called for Malikindjin and said, "Let us fight with spears." Malikindjin fixed a day, "the day after to-morrow I shall come at midday." Malikindjin brought a big bundle of spears and said, "Have you dreamt of me? Did I kill the woman?" and Laindjin said, "Yes. I sent my *tshumbuk* (to investigate). The *tshumbuk* woke me up and told me—'Malikindjin killed her for no reason. The woman had no *ngglambi*. She was not killed by many men. It was only Malikindjin'." Then Malikindjin said, "Come on, let us fight," and Laindjin said, "I am waiting for you. Then we shall fight together."

The two fought hard and their *lanoa* (sister's husbands, etc.) tried to hold them back but they could not. Then Laindjin was wounded in the hand and the *lanoa* managed to hold him.

In compensation[1] for the death of the woman Laindjin paid one mother-of-pearl crescent, one tortoise-shell arm band, two *Conus* shell necklaces and three *Turbo* shells, in a string bag which was hung up in the Wolgem ceremonial house.

This series of stories contains about forty identifications, but such a series is by no means sufficient to give any estimate of the relative frequency and importance of the various types of identification. The series serves only as a set of illustrations of the working of identification in these contexts. It is worth remarking, however, that a rather larger series of stories, properly documented with their kinship setting, would give enough identifications to call attention to the factors which influence the distribution of identification.

[1] It is not clear whether the presentation of homicidal ornaments preceded this compensation or followed it. In other cases, it appears that the homicide pays compensation *before* the relatives give him *tambointsha*.

CHAPTER VI

Structural Analysis of the Wau-laua *Relationship*

W E have seen that the term "mother" is used for mother's brother in the *naven* ceremonies. This use is not confined to this ceremony; it occurs too in initiation, and to some extent in ordinary conversation. Thus certain male natives who regarded me as their classificatory "sister's child" on the basis of a matrilineal name which had been given to me, used occasionally to refer to themselves as my "mothers" when stressing their kinship connection with me. One of them would even say to me: "You are the child that we bore." ("We" here refers to the maternal clan.) At other times informants would refer to the clan, whose matrilineal name I bore, with such a phrase as: "They are your mother."[1]

Thus it is clear that identification of the mother's brother with the mother is very definitely formulated in the terminology, and from the linguistic usage we may expect to find that the relationship with the whole maternal clan is coloured by the relationship to the mother and that the resemblance to a mother is most strongly marked in the case of a *wau*. But we need not expect to find that the *whole* of a mother's brother's behaviour can be described as the behaviour of a male mother.[2] I mentioned above that a man is to some extent identified with his father, and we may expect to find that this identification has some effect on the mother's brother relationship. The *laua* may be to some extent identified with his father, and in this case we shall find the *wau* behaving towards him as if he were the *wau*'s sister's husband. We may therefore expect the *wau*'s behaviour to fall into at least two categories according as he is acting upon the brother-sister

[1] The language has no plural form for nouns so that it is not certain whether the word "mother" in this sentence should be plural or singular.

[2] In this analysis I am much indebted to Radcliffe-Brown's classic paper on the Mother's Brother (*South African Journal of Science*, 1924, p. 542).

identification and treating the *laua* as a mother would treat her child, or upon the father-son identification and treating the *laua* as a man would treat his sister's husband.

The same considerations apply, *mutatis mutandis*, to the behaviour of the *laua*, who may therefore be expected to treat his *wau* both as a mother and as a wife's brother. Lastly it is possible that the *wau-laua* relationship has features which are due to other identifications which we have not considered, or that the relationship may have certain features *sui generis*.

Thus we are led to an experimental analysis of the *wau-laua* relationship under the three following heads:

1. Behaviour which is typical of the relationship between a mother and her child.
2. Behaviour which is typical of the relationship between a wife's brother and a sister's husband.
3. Other details of behaviour which do not fit either of these patterns.

By working with these three categories we shall be able to set out all the available information about the *wau-laua* relationship in a scheme which will show clearly how much or little of a *wau*'s and *laua*'s behaviour is to be regarded as *sui generis* and how much is based upon these identifications.

1. The Iatmul mother's behaviour towards her offspring is simple, uncomplicated by self-contradictory elements such as characterise the patterns of a father's behaviour. Her acts and attitudes may be seen as built up around one central cultural act; she feeds the child, first suckling[1] it and later giving it food. It is evident too from the behaviour of the *wau* that the act of giving food is regarded by the natives as in some sense the essential characteristic of motherhood.

The food so given is a free gift for which the child makes no concrete return, but the mother is anxious lest the child go hungry and she is compensated for her care by her joy in the growth of the child and her pride in its achievements.

---

[1] It may be objected that the suckling of the child is a "natural" and not a "cultural" act. But at least it is an act which is included among the norms of Iatmul culture and which has no doubt been culturally modified in many ways. We are justified in regarding the suckling of the child, from a structural point of view, as one of the *given* facts of Iatmul culture.

The tone of her pride is well illustrated by the incident in the *naven* in which the homicide marches over the prostrate and naked bodies of his "mothers" and they say: "That so small place out of which this big man come!" The attitude is one of humbleness and abnegation of self in pride and joy at the achievement of the offspring. Lastly we have the behaviour of the mother as comforter of the child; she stops its crying and stays beside it in sickness; when the child is too sick to eat, she herself will go without food.[1] The position of the mother may thus be summed up under headings of: food-giving, pride and self-abnegation, comforting.

These aspects of the mother's position are clearly stressed in the behaviour of the *wau*. We see him as food-giver: I have already mentioned that there is a continual stream of ceremonial gifts of food flowing from the *wau* to the *laua*. At small *naven* ceremonies, a fowl, but on more important occasions, a pig, will be presented. We may regard these gifts as analogous to the constant feeding of the child by its mother. It is articulately stated by the natives that the *wau* is anxious lest his *laua* be hungry. A small boy is brought up to believe that unless he is careful how he asks his *wau* even for a single yam, his *wau* will exclaim with exaggerated anxiety, "What! Is my *laua* hungry?" and jump to his feet and run to kill his pig and present it to the boy.

We see the *wau* combining self-abnegation with pride in the boy's achievements: The attitude of vicarious pride is daily exemplified by the *wau's* hailing the boy with a series of names of outstanding totems of the maternal clan, and more dramatically, whenever the *laua* makes a conspicuous public appearance, e.g. in debate, by the *wau's* throwing lime over him. Again the attitude of pride is dramatically combined with modesty and self-abnegation in the *naven* ceremonies when the *wau* wears filthy female garments, shaming himself to express his pride in his *laua*. Indeed, even at this stage of our enquiry we may conclude that a great part of

---

[1] I have no evidence that this detail of a mother's behaviour is carried over into the behaviour of the *wau*, but in a certain stage of the ceremonies of initiation, the term *nyame* (mother) is applied to the elder age grade, and during this period the members of this grade observe food taboos with the novice.

the *wau*'s behaviour in *naven* is an exaggerated statement of this particular aspect of maternity, the mother's self-abnegation combined with pride in the boy. But many details of the *naven* still remain to be accounted for—especially the element of burlesque in the *wau*'s behaviour and the *wau*'s offering of his buttocks to *laua*.

We see the *wau* as a comforter in the ceremonies of initiation. At this time he is called "mother" and he actually behaves as a comforting mother to the little boy during the grim first day of initiation when the child is separated from his real mother and undergoing scarification. The *wau*'s first act in initiation is to lift his *laua* pick-a-back—the manner in which mothers often carry their children—and carry him to the upturned canoe on which the scarification is to be performed (cf. Plate VB).

During the operation the *wau* sits on the canoe, and the little boy sits at first in his *wau*'s lap while the incisions are made around his breasts. After that the boy lies with his head in his *wau*'s lap and his back exposed to the knives of the initiators. The cutting of the back is much more painful than the preliminary cutting of crescents round the breasts and the little boy screams and clutches as hard as he can with his arms round his "mother's" waist. When he screams the *wau* will say, "Don't cry," and grasps him, holding the boy still and at the same time responding to his frantic clutches. When the cutting is at last finished, the *wau* again takes the child pick-a-back and carries him to a pond where the blood is washed off. Then the *wau* carries him back to the ceremonial house where he soothes the boy's cuts with oil gently applied with a feather (cf. Plates X, XI).

Turning now to the reciprocal behaviour of the boy towards his *wau* we can trace again points in which it is modelled on the type of his behaviour to his mother. In the attitude towards the mother there is a strong emphasis on loyalty: I have already published[1] an incident in which a son interfered in a quarrel between his mother and her second husband. The husband had lost his temper and beaten the woman seriously with fire sticks. Her son entered his stepfather's

---

[1] *Oceania*, 1932, p. 286.

house and took her away and lodged her in his own house for several days. This same emphasis on loyalty is brought out in the behaviour towards mother's brother. Whenever a quarrel starts in which his *wau* is involved it is said that the boy should take his *wau*'s side, even if the quarrel is one in which his father is opposed to his mother's brother.

The same loyalty is emphasised in matters connected with esoterica—spells, mythology, etc. (cf. p. 37).

As a complement to the *wau*'s pride in the *laua*'s achievements, boasting is developed in the latter. He will swagger in his *wau*'s presence and will say with great gestures: "I am the big man of this village!" "I am the house-post which supports this village!" But at the same time he will not swagger overmuch, and especially he must not insult his *wau*, lest the latter jump up and rub his buttocks on the *laua*'s shin. In the easier relationship between a boy and his mother, this boasting behaviour is not conspicuous, and its development in the *wau-laua* relationship is probably correlated with the generally exaggerated character of the *wau*'s behaviour.

In this exaggeration, the behaviour of the *wau* differs from that of a mother. She gives food easily to her child, but when the *wau* gives food he either presents pigs in a ritual context, or jumps up, and, with a histrionic gesture, kills a pig in response to the *laua*'s demand for a yam. The mother carried the small child on her shoulders because that was a convenient way of carrying it; when the *wau* carries the boy pick-a-back, the act is an incident in the ceremonial of initiation when the boy is grown and well able to walk. The mother may smile easily at her boy's achievement, but the *wau* is impelled at least to throw lime over him. In every instance the behaviour of the *wau* is an exaggerated and dramatic version of that of the mother. The formulation of the social identity between a man and his sister, between *wau* and mother, may be a sufficient explanation or description of the resemblances in the behaviour of these two relatives, but the elements of burlesque and exaggeration in the *wau*'s behaviour still remain to be explained.

2. We may now consider the patterns of behaviour typical of the relationship between a man and his sister's husband

and the extent to which these patterns are adopted by the
wau. The chief characteristic of the brother-in-law relationship among the Iatmul is a mutual ambivalence about the
fact that one man has given his sister to be the wife of the
other. A feeling of mutual[1] indebtedness and distrust is
culturally expected, and a considerable bride price is given
to the brothers of the bride by the husband. Although this
transaction is phrased as *waingga* (purchase), a considerable
quantity of valuables are loaded on to the wife's body by her
brothers when she goes to her new home and these valuables
are a return presentation to the husband. The natives say
quite articulately that the purpose of the bride price is "to
prevent the wife's relatives from using sorcery against the
husband", and they say that this resort to sorcery will be
staved off only for a limited time by even the biggest bride
price. The sense of indebtedness therefore remains and the
wife's relatives have always the right to call on the husband
for help in any task, like house-building, for which a crowd
of manual labourers is necessary. When the task is completed
the wife's people will stand a small feast for the labourers
or they will at least distribute coconuts to them. This largess
is of the nature of a complimentary presentation and is
usually quite inadequate as a payment for work done.

But the brother-in-law relationship is ambivalent; it is
characterised not solely by indebtedness and distrust but also
by co-operation. In many cases the bride's brother takes an
active part in the arrangements for the marriage. He may
assist his friend, the future husband, in his courtship by
putting magic in his sister's fireplace to cause her to fall in
love; and finally he may lead her by night from their house
to that of the bridegroom. On arriving there, the man and
his sister are ceremonially received by the future husband.
The girl's brother is made to sit on a stool while the future
husband puts oil on his face, cuts his hair and finally puts
valuables on his shoulder.[2] Next morning the village wakes
to find the marriage a *fait accompli*.

[1] Cf. the quarrelling between wife's fathers and husband's fathers,
described on p. 105; Kwoshimba *versus* Kili-mali.

[2] Note that the ceremonial acts performed by the future husband are
the same as those performed by an offender in making amends to the man
he has offended (cf. p. 98).

In many cases too the opposition between brothers-in-law is treated lightly and used as a subject for joking. If a man is idly watching his *tawontu* doing some job, the latter will say merrily enough, "Watching your *tawontu*? I am not that sort of man. You should see me help my *tawontu*s." It is probable too that the opposition between brothers-in-law is reduced to a minimum in those cases in which the relationship between them is symmetrical, based upon an exchange of sisters. But no native statement was collected to this effect.

Even without this exchange there is an insistence that a man should be loyal to his wife's brother—a feature in which this relationship resembles that with the mother and the mother's clan. Indeed, we find here the same type of precept which we noted in the relationship to the mother's clan—in this case, that a man should support his wife's people in all their quarrels, even against his own clan. In this respect then, father and son are alike in owing absolute allegiance to the same clan, and the insistence upon loyalty to the *wau* may be phrased as consistent both with identification of that relative with the boy's mother and with the identification of the boy with his father.

When a quarrel arises between a man's maternal clan and the clan into which he has married, he should either stand up between the disputants and stop the fight, or hold his peace. In general, these precepts are to be regarded as statements of what a man ought to do rather than as statements of what is usually done. But I have on several occasions seen men intervene in brawls in the ceremonial house. Unfortunately I never examined the kinship ties between them and the disputants. At such times I was myself anxious lest some serious damage might be done which would call down the wrath of the Government upon my village and perhaps deprive me of the services of valuable informants. I presumed that the motives of the peacemakers were a product of modern culture contact and fear of the Government. It never occurred to me, even though I knew the precepts, that such kill-joy conduct could be culturally normal to the Iatmul.

Besides the insistence upon loyalty, other aspects of the

brother-in-law relationship are taken over into the *wau-laua* relationship. A husband gives valuables and labour and receives a wife—and these gifts are tied together by a sense of indebtedness. This uneasy economic aspect of the brother-in-law relationship has been taken over into the pattern of the *wau-laua* relationship. We saw above that the *wau* makes considerable presentations of food—especially pigs and fowls —to the sister's son. But these gifts—unlike the corresponding food gifts of the mother—prompt the boy to a sense of indebtedness, and he makes in return presentations of shell valuables to his *wau*. Actually in many cases the boy has no wealth of his own with which to recompense his *wau* and it is the father of the boy who provides the shells for the presentation, a fact which illustrates the identification of sister's husband with sister's child most vividly in the present context. Stated from the *wau*'s point of view: he presents a pig to his sister's son and his sister's son and his sister's husband combine their wealth to make a return present of valuables.

Lastly we may consider the extraordinary conduct of the *wau* in presenting his buttocks to his *laua*. Such conduct is of course not characteristic of a mother, but I have a casual mention in mythology of a man who rubbed his buttocks on the leg of the man who was marrying his sister. If we bear in mind the identification of a man with his sister, this conduct is comprehensible—at least from a structural point of view. The man expressed his relationship to his sister's husband by ritually making a sexual gesture in which he identified himself with his sister. In the case of the *wau*, the position is more complicated, but his behaviour becomes perfectly logical if we think of the two identifications upon which it is based. For the sake of clarity we may state these identifications as if the mother's brother were speaking, viz. "I am my sister" and "My nephew is my sister's husband." If now we consider these two identifications simultaneously it is perfectly "logical" for the *wau* to offer himself sexually to the boy,[1] because he is the boy's wife.

[1] I should perhaps stress the fact that this interpretation of the *wau*'s gesture is my own and is not based upon native statements. It is conceivable that the gesture may be symbolic of the *birth* of *laua* from *wau*, a further stressing of the maternal aspects of the *wau*'s position.

This last piece of analysis may seem to be fanciful, but the evidence goes to show that there is some such process of logical symbolism underlying this piece of ritual. Upon this hypothesis the *wau*'s exclamation: "*Lan men to!*" "Husband thou indeed!"[1] becomes immediately understandable. Lastly the fact that the *wau*'s sexual gesture is said to act as a strong stimulus to make the *laua* run to get valuables in order to present them to his *wau* is explained if this presentation is really in some sort a bride price.

But although the structural position of the *wau* is such that his actions in the *naven* ceremony can be described as its logical development, we still do not know why the culture should have followed up this line of logic. Why should the culture have followed these various identifications to their ultimate *reductio ad absurdum*, a *wau* offering himself sexually to his *laua*? Such formulated identifications between a man and his sister and between a man and his father can be traced in many primitive communities, but even the simpler elements of the *naven* ritual—the *wau* dressing as a woman, etc.—are comparatively rare in such societies; and such a phenomenon as the *wau*'s sexual gesture is perhaps only to be found in the Sepik area. We must therefore conclude that while the structural position lays down possible lines along which the culture *may* develop, the existence of such lines does not explain why the culture should select them for emphasis. We have demonstrated the existence of the lines but we have still to demonstrate the "motive force" which has caused the culture to run along them.

3. Lastly we have to consider the various points in the *wau*'s behaviour which cannot be described by saying either that the *wau* is a male mother of the *laua*, or that the *laua* is the sister's husband of the *wau*.

---

[1] The evidence from this exclamation is, however, not as strong as its literal translation would indicate, since the words, *lan men to*, are also constantly used as an expression of submissiveness, a demand for compassion and a cry of surrender. The phrase can even be used in addressing a female oppressor and is then modified to "*lain nyin to*". (*Nyin* feminine, second person, singular pronoun. The modification from *lan* to *lain* is, I think, not expressive of change of gender but merely "euphonic".)

(*a*) The *wau* will not take fire from the *laua*'s house. If the fire goes out in his own house, the *wau* will, it is said, go and light his cigarette at the *laua*'s fire and from the cigarette will light his own fire, but he will not take a burning fire stick from the *laua*'s fire. This type of avoidance is more typical of the boy's father, who, as I have mentioned, will never under any circumstances eat of food collected by his son—fire and food being closely analogous in Iatmul thought.

It would appear that the *wau*'s avoidance of taking fire from his *laua*'s house is perhaps structurally described by an identification of the *wau* with the *laua*'s father; and this is the more probable since the complete series of taboos on eating food collected by the boy is applied to the father's *tambinien*. The *tambinien* is a partner and the behaviour between *tambiniens* is modelled upon the same general lines as the reciprocal co operation between men who have exchanged sisters. Thus it is not surprising to find that a similar identification obtains to some extent between simple brothers-in-law.

(*b*) The *wau* resembles a father in another point; like a father he avoids entering upon crudely commercial transactions with his *laua*. But it is not perfectly clear that this resemblance between the two relatives is due to an identification between them and not to some coincidence whereby different causes operate upon the two relatives to induce superficially similar behaviour. In the case of a father, the avoidance of commercial relations with the son is, no doubt, a part of the general avoidance of profiting at the son's expense. In the case of the *wau*, I was told that the avoidance is specifically connected with the fact that this relative has received a part of the bride price paid for the *laua*'s mother; but I do not know why this fact should make it uncomfortable or unfitting for a *wau* to trade with his *laua*; nor do I know whether commercial transactions occur between brothers-in-law.

(*c*) Another usage may doubtfully be referred to the same identification: when a *laua* performs some service for his *wau* he is spoken of as the latter's "dog". I do not know exactly what details of the relationship between *wau* and *laua* are here equated with what details of the relationship between a man and his dog; but it is worth remarking that the owner of a dog is ordinarily spoken of as its "father".

(*d*) I have already described the series of details of behaviour which seem to equate the *laua* with the ancestors of the maternal clan and suggested that this equation is based on a similar orientation of pride towards ancestors, *laua*s and achievement.

(*e*) The behaviour of the *wau* is *exaggerated*. It is not a mere copy of that of a mother or brother-in-law, but is a sort of burlesque of the behaviour of these relatives. This element of burlesque must for the present be described as peculiar to the *wau* relationship and not as derived from any pattern of identifications.

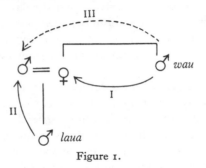

Figure 1.

From the foregoing analysis it follows that we may sum up the principal characteristics of the *wau*'s position by means of a diagram (Fig. 1). In this diagram the genealogical position of *wau* and *laua* is shown and three arrows are drawn representing the three identifications on which the behaviour of *wau* and *laua* is based. Arrow I marks the identification of *wau* with his sister, according to which *wau* will behave as a mother, and the *laua* will behave as a son towards a mother. Arrow II marks the identification of the *laua* with his father according to which *wau* will treat *laua* as if he were a sister's husband and *laua* will treat *wau* as if he were a wife's brother. The combination of identifications I and II is the basis for the *wau*'s treatment of *laua* as if he were a wife.

Arrow III marks the identification between *wau* and *laua*'s father, according to which *wau* will behave as if he were *laua*'s father, and *laua* will behave as if he were *wau*'s son. Since this last identification is much more problematical than the other two and since I have no record of any behaviour of the *laua*, which would fit this pattern, I have indicated this identification with only a dotted line on the diagram.

Our experiment has shown that almost the whole of the *wau*'s cultural behaviour can be analysed in terms of two identifications. Of the details which remain unaccounted for, some can be referred to a third identification but a few still defy analysis in these terms. We still do not know why the *wau*'s behaviour tends to be exaggerated and comical, why the *laua* is regarded as the ancestors of the maternal clan, and why this culture has followed the logic of its identifications to such extreme conclusions. These problems must be left for solution in terms of aspects of Iatmul culture other than the purely structural.

CHAPTER VII

The Sociology of Naven

Hitherto we have confined ourselves rigidly to what I have called the Structural point of view. We have put ourselves in the place of a hypothetical intellect inside the culture and have shown that to such an intellect the behaviour of *wau* and *laua* in the *naven* is in some sense logically consistent with the other given facts of the Iatmul culture. But in the present section we shall consider the *naven* from an entirely different point of view, that of an observer outside the culture and interested in the integration and disintegration of Iatmul communities. We shall consider only those aspects of *naven* which are relevant to the well-being of that abstract unity, the Society as a whole. This shift of viewpoint involves a complete re-orientation towards the culture and a complete re-phrasing of our problems.

The fundamental problem with which we are now concerned is: Does the *naven* have any effects on the integration of Society? This problem I shall modify by making an assumption of a type not unfamiliar to anthropologists. It is evident that the *naven* ceremony is an expression and a stressing of the kinship link between the *wau* and *laua* concerned, and from this I shall assume that by this stressing the link is strengthened.[1] In the light of this assumption, the problem of the sociological functions of *naven* becomes: What effects has the strengthening of this link upon the integration of the Society as a whole?

To answer this question we shall have to ask many subsidiary questions, and the form of these will contrast sharply with the form of the subsidiary questions which we asked in our structural enquiry.

---

[1] This is not a sociological assumption but refers to a process of psychological conditioning of the individuals concerned. The question of its validity should therefore be examined not in the present chapter but in a chapter devoted to the study of character formation among the Iatmul, a subject which I did not investigate in the field.

At the outset it appears that a sociological enquiry should be documented with statistics, a method of treatment which would have been excessively awkward in the study of cultural structure. At the most, the only statistical questions which could have been relevant to our structural enquiry would have taken the form: "How many details of behaviour imply so-and-so?" But in the present section, the whole phrasing of the problem being different, the type of statistical question which we shall ask will take the general form: "How many individuals do so-and-so?" a form of question to which answers could have been collected in the field.[1]

If we consider a simple example from our own culture this difference in the framing of the statistical questions will be clear: Fathers beat their children. This fact could be structurally investigated and it could be shown that the beating of children was consistent with other details of the father-child relationship; and this consistency would be equally demonstrable though our note-books only contained reference to one case of child-beating. But if we are investigating the same fact from a sociological point of view we shall need statistics. We may guess that the beating of children is important for the general maintenance of order in Society. But to demonstrate this function conclusively we need statistics which will answer questions of the general type, "What percentage of such and such individuals do so-and-so?" Specifically in the present instance we shall want to know:

1. What percentage of fathers beat their children?
2. What percentage of beaten children in later life go to prison for riotous behaviour?
3. What percentage of unbeaten children later go to prison? Etc. etc.

From this example we may deduce that the emphasis of the sociological questions is likely to be, not "How many details of cultural behaviour...?" but rather "How many individuals...?" Further, it appears that we must expect

---

[1] Unfortunately I have no statistics and took no random samples. The conclusions of this chapter are therefore unproven. The chapter itself is only included for the sake of giving an illustration of the problems and methods of approach of sociology in the strict sense of the word.

in a sociological enquiry to be concerned with classes of individuals—"beaten children", "unbeaten children", "fathers who beat", etc. In fact, it will be our business to classify, not details of behaviour according to their implicit assumptions, but individuals according to their behaviour.

Returning to the problem of *naven* which I stated above in terms of the sociological importance of the link between *wau* and *laua*, one fact now stands out as likely to be a clue to the position: it is the *classificatory wau*s who perform *naven*.[1]

The relevance of this fact becomes clear when we examine the marriage system of the Iatmul and the patterns of behaviour between groups of individuals who are linked together by marriages in past generations.

In many primitive societies we find marriage regulated in a positive manner by kinship. Not only are there prohibitions of marriage with certain relatives, but also positive injunctions which constrain a man to choose as wives women who are related to him in some specific manner, real or classificatory. Under such systems the links of affinal allegiance and indebtedness are regularly renewed in succeeding generations. But in Iatmul society this is not the case. (Lacking statistics I must here substitute a description of the marriage system in formal terms.) The culture does, it is true, contain a great many formulations which would regulate marriage in a positive manner if they were consistently obeyed. We find for example such statements as:

1. "A woman should climb the same ladder that her father's father's sister climbed", i.e. she should enter, as a bride, the house which her father's father's sister entered. This is a way of stating that a woman should marry her father's father's sister's son's son (or reciprocally that a man should marry

---

[1] This is certainly the case in Palimbai and Kankanamun. But the position in Mindimbit is probably rather different. In the only *naven* which I studied in detail in the latter place, the *naven* for a girl who had caught a fish, it was the own *wau*s who took part. In Palimbai the *naven* for a boy who had made a new canoe was performed by classificatory *wau*s and it was in Palimbai that I was told that a man has two sorts of *tawontu*, those who receive a share of the bride price of the man's wife and those who will celebrate *naven* for the man's children. It is possible that the sociological functions of *naven* in Mindimbit differ profoundly from those I have suggested for Palimbai and Kankanamun.

his father's mother's brother's son's daughter, or *iai*.[1] Such
a marriage rule if carried out consistently would lead to a
repetitive system and the affinal links would then be renewed
in every alternate generation. But the rule is not so carried
out and we find other conflicting formulations.

2. "The daughter goes as payment for the mother." This
is a way of stating that a man marries his father's sister's
daughter; and this practice again would lead if it were consistently obeyed to a repetitive system,[2] and one which,
though conflicting with the *iai* marriage mentioned above,
is still generically connected with it in a curious way.

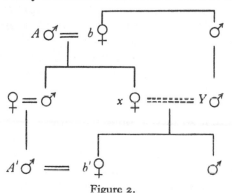

Figure 2.

This generic relationship between the two systems is illustrated in the diagram, Fig. 2. The marriage between *A'♂*

---

[1] Cf. *Oceania*, 1932, p. 263. There is a second cliché which defines
iai marriage: "*Laua*'s son will marry *wau*'s daughter." This refers to
marriage of a man with his father's mother's brother's daughter, a relative
who is also called *iai*. This cliché was collected both in Mindimbit and
in Palimbai, while the cliché referring to the ladder was collected only
in Kankanamun. It is possible that in Mindimbit the sociological function
of *naven* is the promotion of marriage between *laua*'s son and *wau*'s
daughter. Thus the *naven* becomes a "device" not for perpetuating
affinal links which would not otherwise be repeated in the community,
but for bringing about their repetition. I have, however, no statistics or
native statements which would support this theory.

[2] Cf. Dr Fortune, on Kinship Structure, *Oceania*, 1933, pp. 1–9.
Among the Iatmul, however, this type of exchange is not strictly
confined to the regulation of marriage. In certain cases, when the offspring of the marriage are male instead of female, one of the sons will
be sent, while he is still a baby, for *adoption* into the family and clan of
the man who gave his sister to the father for wife, i.e. the boy is adopted
by his mother's brother, to whom he goes as payment for his mother.

and $b'♀$ is an *iai* marriage which repeats the original marriage $(A ♂ = b ♀)$ in the grandparent generation. The difference between the *iai* system and the father's sister's daughter system turns on whether the marriage, $x ♀ = Y♂$, occurs in the intervening middle generation. If this marriage occurs, the whole system is converted into one based upon marriage with the father's sister's daughter, and the final marriage, $A'♂ = b'♀$, becomes a marriage of a man with his father's sister's daughter. The introduction of this intervening marriage is linked with a new formulation:

3. "Women should be exchanged." This is more usually stated by the natives in reference to sister exchange—a man giving his sister as wife to the man whose sister he himself marries. Such a formulation, stated in this way in terms of sister exchange, of course conflicts with either *iai* marriage or father's sister's daughter marriage. But the fundamental concept of the exchange of women is common to both sister exchange and marriage with father's sister's daughter.[1]

Thus Iatmul society is built upon three formulations in regard to marriage which though they conflict are still inter-related in a curious logical manner. The father's sister's daughter marriage is comparable with the *iai* marriage since both fit into the same pattern of repetitions in alternate generations, and the sister exchange is comparable with father's sister's daughter marriage since both depend upon the exchange of women.

This discussion of the marriage rules was introduced to show that the society is without any regular repetitive system whereby the affinal links might be regularly renewed. I have elaborated the description somewhat more than is necessary in the present context because the cultural details here given

[1] It is quite likely that the marriage with father's sister's daughter has been evolved in the culture through interaction of the two systems, *iai* marriage and sister exchange. The whole kinship vocabulary would point to the *iai* system as the older, and the concept of exchange of women may well have been adopted from neighbouring peoples. In this connection it is interesting that one of the young men whom I used as an informant had a strong impression that the correct term for father's sister's daughter was *iai*. But after we had discussed the matter in detail, he was uncertain and consulted one of the older men who stated definitely that *na* was the correct term for this relative.

will also be relevant in a later chapter. We shall see that this society constantly dovetails together ideas which are incompatible and that there is perhaps something in common between the muddled logic which underlies the *naven* ritual and that on which the rules of marriage are based. In the discussion of the totemic system, we shall see (p. 128) the capacity of the natives to take a pride in the schematic quality of their system, although actually that system is riddled with incompatibilities and fraud. In their marriage rules too the natives take a pride and they look down on their neighbours as "dogs and pigs who mate at random".

This gibe is singularly inappropriate in the mouths of the Iatmul, since not only have they three positive marriage rules which conflict one with another, a circumstance which would in itself introduce a random element into their mating, but the people do not adhere even to their negative rules. These negative rules are very vague, but there is a strong feeling against marrying own sister and I never collected a case of marriage of this kind. The next strongest ban is that upon marriage with any of the relatives called *naisagut* (e.g. wife's mother), and I found a case in which a man[1] had sufficient standing in the community to marry his own wife's own mother and this while the wife was still alive and married to him. He was a great sorcerer and at the same time a great debater and fighter. It was nobody's business to say him nay in this individualistic culture. Such marriages are rare—even among the Iatmul—but when we turn to the less stringent prohibitions, as of marriage with classificatory "sisters", women of own clan with whom genealogical connections can easily be traced—we find that such women are sometimes taken as wives. When I enquired into these marriages, I was generally answered with a cliché: "She is a fine woman so they married her inside the clan lest some other clan take her." The Iatmul even rather approve of these endogamous marriages and say that from endogamy are produced long and widely ramifying lines of descendants—a belief, the sociological and cultural significance of which is not at all clear to me.

Occasionally, however, the members of a large clan,

---

[1] This was Mali-kindjin, of whose character a sketch is given on p. 164.

apparently realising that they are becoming too endogamous, proceed to formulate a statement that one half of the clan belongs to the bows of the clan war canoe and the other half belongs to the stern. Henceforth these two groups are regarded as separate clans for purposes of marriage and what is, in a sense, really endogamy is satisfactorily phrased as exogamy.

Marriages with classificatory "mothers" are also not uncommon.

Lastly, besides these endogamous marriages, there are many marriages with outside groups—with women captured in war, or sent as peace offerings, women met on trading expeditions, etc. So that we may sum up the marriage system by saying that in practice marriage occurs very nearly at random; and when we consider that the villages are very large, with a population ranging from two hundred to a thousand, it seems unlikely that an important affinal link will be perpetuated by analogous marriages in future generations. If therefore these old affinal links are necessary for the integration of the community, some means must be found of diagrammatically stressing them, a function performed by *naven*.

It remains for us to consider whether these linkings are really important and how they are used in the integration of society. In this connection the most significant fact is that the formulations of proper behaviour towards affinal relatives can be applied in this culture not only to own wife's own relatives and to own sister's husband's relatives, but can be extended to a whole series of relatives who are grouped around this central nucleus in a classificatory manner. Such fundamental principles as the identification of siblings, the identification of a man with his father and with his father's father, and the grouping of whole clans as single units, are applied in this extension of affinal relationships. But it would not be true to say that the affinal linkages due to every marriage are extended indiscriminately in every direction according to this scheme of identifications.

The formulations of proper behaviour contain such precepts as that a man should be loyal to his wife's relatives and should help them in the work of building their houses and their ceremonial houses; he should help them pull their canoe logs, clear their gardens, and he should come to their assistance in all their quarrels. If all affinal linkages were observed no one would be able to quarrel with anyone else inside the community, and everybody would have to go everywhere and do everything with everybody else, since the genealogical links are actually ubiquitous.

One need only spend a very few days in a Iatmul village to realise that there is no such thorough integration of the population as this complete and indiscriminate extension of affinal ties and duties would imply. Quarrels are not rare and whenever a large group of workers is required for some great labour such as house-building or repairing the ceremonial house, it is difficult to get the men together and difficult to get them to co-operate. But in the end the groups can be got to work and such co-operation as is achieved is certainly made possible by the insistence upon affinal ties which extend in a perfectly definite manner along certain chosen lines of the classificatory system.

There are two main types of extension of the affinal system and various collective terms are applied to the extended groups so formed. The two commonest terms are *lanoa nampa* (literally "husband people") and *laua nyanggu* ("sister's-child children"—the word "children" in this context being no more than a synonym for "people").

The term *lanoa nampa* is strictly most applicable to the group formed by the first type of extension of the affinal system. This is the group of persons whose affinal relationship results from some definite *contemporary* marriage, and it contains such relatives as: sister's husband (*lan-ndo*), sister's husband's brother (*lan-ndo*), sister's husband's father (*laua*) and sister's husband's son (*laua*). Thus, when the relationship depends upon a recent marriage, the term *lan* is extended to include various *laua*s. (The reciprocal collective term for relatives of the wife is *towa-naisagut.*)

In the second type of extension of the affinal system, the relatedness of the groups is traceable to a *past* marriage, but the reality of the connection between the groups is demonstrated by the *naven* system. In this type of extension, the more correct term for the group, which contains the relatives of the former husband, is *laua nyanggu* (though the term *lanoa nampa* is sometimes used). Reciprocally, the group containing the relatives of the former wife is called *wau nyame nampa*. Thus the term *laua* is here extended to include classificatory sister's husbands, the reverse of the usage mentioned above in which the term *lan* is extended to include sister's children.

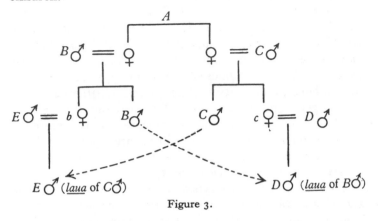

Figure 3.

The natives are perfectly conscious of the genealogical process upon which the formation of this *laua* group of relatives depends, namely, the perpetuation of past marriages. There is a native proverb which says: "Legs of a *Caryota*,[1] legs of a *Pandanus*; women hither, women thither." The *Pandanus* and the *Caryota* are alike in having visible aerial roots diverging downwards to the ground from the trunk; and the meaning of the proverb is that diverse groups are tied together by affinal bonds due to past marriages of pairs of sisters. The genealogical pattern of the spreading of these links is represented in the diagram, Fig. 3.

Two sisters of clan *A* get married. One of them marries into clan *B* and the other marries into clan *C*. Their offspring

[1] This identification is uncertain. I should rather have said "a species of palm which looks like a *Caryota*" but this phrase could hardly be inserted into the proverb.

are regarded as classificatory siblings although the children
of the one woman are of clan *B* and those of the other of
clan *C*. In this filial generation, the girls will again marry
into other clans, e.g. the girl of clan *B* marries a man of
clan *E*, and the girl of clan *C* marries a man of clan *D*. The
grandchildren from the original marriages will again regard
themselves as classificatory siblings. The members of what
is now the middle generation will become classificatory
"fathers", "mothers", "mother's brothers" and "father's
sisters" of these children: so that owing to the original marriage of two sisters of clan *A* it has come about that a man in
clan *B* has a *laua* in clan *D* and a man in clan *C* has a *laua*
in clan *E*.

But though the term *laua nyanggu* may be applied to all
these classificatory *laua*s and even extended to apply to their
fathers and other members of their clans, a man will take
steps to mark out his relationship to the *laua* groups whose
allegiance he especially desires[1] and he will demonstrate that
he really is a *wau* by giving, to the baby *laua*, a name of
which the termination means "mask". This gift of a name
is accompanied by the gift of a coconut and it is followed in
later years by the whole *naven* sequence. The *wau* will make
it his business to remind the *laua* constantly of his relationship; whenever he meets the boy he will address him by the
name which he gave him, he will exclaim at the *laua*'s achievements and he will present the *laua* with meat. The *laua* will
give return presents of valuables to the *wau*, and will come
and work whenever the *wau* is gathering all his *lanoa nampa*
and *lauu nyunggu* for some heavy work in house-building or
clearing a garden.

These two types of affinal relationship were even more
clearly defined by one informant, in Palimbai, who explained
to me that he had two sorts of *tawontu* (wife's brother). Of
these relatives, one kind were "the men who had received
a part of the bride price he gave for his wife", i.e. the members
of his wife's clan of her generation. The other series were

---

[1] Unfortunately I did not collect any material bearing on the circumstances likely to influence a *wau* in his choice of one classificatory *laua*
rather than another for this recognition.

" *tawontu* who would celebrate *naven* for his children ". Thus
the bride price serves as the defining factor in the relationship with one group of affines, and the system of *naven* and
names given by the maternal clan serves as the defining factor
in the other case.[1] By the combination of these two[2] systems,
the community is so knit together that it can accomplish
great architectural works and warfare.

It may be asked why the culture should preserve these old
affinal linkages by stressing the resulting classificatory *wau-
laua* relationships rather than by stressing the corresponding
classificatory *lando-tawontu* relationships—and this in spite
of the fact that the latter would seem, to a European, to be
the simpler and more direct. The answer to this question is,
I think, to be found in the fact that the actual system fits
with the remainder of the culture, e.g. with the naming
system in which *wau*s give names to their *laua*s. In fact, the
answers to questions of this type are to be sought not in
sociological terms but rather in terms of the cultural structure.[3]

The reality of the integrating effect of *naven* might still
be doubtful if the Iatmul communities were limited in size
by some factor other than the weakness of their internal
integration. It is conceivable, for example, that a community
whose size was limited by its physical environment would
never reach that size at which every contributory integrating
factor becomes relevant. But actually, in Iatmul culture, it is

---

[1] The informant who defined this position was, I think, somewhat
more definite in his statements than the actual facts of the culture justify.
In many cases a child's own mother's brother will give it a name; and it is
likely that there is some variation in these respects from one village to
another.

[2] There is a third type of extension of the affinal system which is
important in the integration of ceremonial and architectural undertakings
in the ceremonial house. This is the relationship between certain pairs
of clans which regard each other reciprocally as *lanoa nampa* or *laua
nyanggu*—both terms are used—and who do a great deal of work for each
other. This relationship is reciprocal between clans and apparently depends
not upon any particular present or past marriage, but upon a tradition
that the women of the one clan often marry the men of the other and
vice versa. The relationship finds ritual expression on various occasions,
especially in mortuary feasts, but is never, I think, marked by the *naven*
system. This clan relationship is practically without effect outside the
ceremonial house.

[3] Cf. Epilogue, p. 269, where another type of answer to this question
is suggested.

clear that the factor which limits the size of the villages is the weakness of their internal cohesion. The larger villages are continually on the point of fission, and the fissions which have taken place in the past are always ascribed to quarrels which have split the parent community. More than this, these fissions when they occur invariably follow the lines of the patrilineal groups—a clan, a phratry or a moiety splitting off from the parent community and thereby rending the system of affinal linkages. From this pattern of cleavage it is evident that the patrilineal linkages are stronger than the affinal, the latter presenting as it were a plane of weakness of the community. Under these circumstances the importance of any factor which strengthens the affinal links becomes apparent and we are justified in saying that the villages could not be as big as they are if it were not for *naven* ceremonies or some analogous phenomenon.

The phrase, "some analogous phenomenon", might well denote a mechanism entirely different from the *naven*, one even which did not rely for its effect upon the stressing of the affinal links. The integration of society might for example be built up by the attachment of all the individuals to some chief, their allegiance to whom would then be strengthened by ceremonial of some quite other pattern. Equally, the same sociological function might be achieved by any of the mechanisms which enforce codified law.[1] Indeed the sociological position of *naven* is hardly comprehensible to a European until he has some picture of how a society works without either officials or codes of law.

The Iatmul are, fundamentally, a people without *law*. By

---

[1] Malinowski has stated in the Introduction to Hogbin's *Law and Order in Polynesia* that an analogue of our own "law" may be found in all cultures, and this is no doubt true if we take a sufficiently abstract view of the matter. But we should not lose sight of the fact that the *only* resemblance between the European legal system and such a system as that of the Iatmul lies in their widest sociological functions. We may, for example, say that a system of codified law, or a strong chieftainship, performs in higher cultures many of the functions which vengeance performs among the Iatmul. But in a functional study we cannot afford to ignore the mechanisms involved—equating jellyfish with fish because they both swim. To the writer it seems that there is a very profound contrast between peripherally oriented systems such as that of the Iatmul and centripetal systems such as those of Western Europe.

this I do not mean that they have not customs and sanctions, but that they have no codified law and no established authority which might impose sanctions *ex officio* or in the name of the community as a whole. It is a general principle—though one not stated as a generalisation by the natives themselves—that no considerable sanction, i.e. no fine, damage to property or physical injury, is ever imposed upon an individual by the group (clan, moiety, initiatory group, or village) of which the individual is himself a member; neither is any considerable sanction ever imposed within the group by a superior authority representing that group or by any outstanding man in the group. In general such internal sanctions confine themselves to disapproval, insult and abuse.[1]

In the absence of such sanctions, internal or imposed from above, the ordering of Iatmul society is almost entirely dependent upon what we may call external or lateral sanctions. The tendency of the people is to phrase every offence as an offence *against somebody* and to leave the business of inflicting sanctions to the offended people.

In its simplest forms the system works thus: *A* offends *B*. *B* avenges himself with weapons or sorcery. Alternatively *B* may produce such a show of anger and power or such a shower of abuse that *A*, either frightened or ashamed, will pay some sort of compensation in shell valuables, pigs or areca nut. *B* will then formally accept this compensation, either stepping on the pig, standing while *A* puts the bunch of areca nuts on his shoulder, or permitting *A* to cut his hair, oil his face and otherwise decorate him. In some cases, *B* will finally make a presentation of some sort to *A*.

This phrasing of the matter in individualistic terms is, as

[1] The only exceptions to this rule are: (*a*) Cases in which an outstanding member of the group will *recommend* the offender to compensate an offended person outside the group. Such recommendations may be angry and vociferous but they remain mere recommendations and are in no sense a decision of the dispute by constituted authority. The outstanding man speaks as one who is concerned in the dispute and who may be seriously involved if the quarrel grows. He is not an impartial outsider. (*b*) Cases inside the family group. Here the husband has authority over his wives; if two wives are jealous he may beat them both. Father has authority over son and elder brother over younger brother. But this pattern is confined to the family itself and is not repeated in the larger groups or in the society as a whole.

we have already seen in the chapter on "Sorcery and Vengeance", to some extent obscured by the phenomena of identification whereby, when *A* offends *B*, the resulting quarrel is taken up by their relatives, fellow clan members, or members of the same initiatory groups. But always the offender expects his own people to be active on his side and the quarrel remains one between two *peripheral* groups, never taking the *centripetal* form which we might phrase as "Rex or State *versus* So-and-so". Such a form is indeed precluded by the considerable feeling which exists in the Iatmul community against any participation of outside persons in quarrels which do not concern them.[1]

The whole mechanism of Iatmul sanctions differs so profoundly from our own, that I must illustrate it with examples:

1. My informant, Tshava, of Kankanamun, was a man of about thirty-five. The following incident occurred when he was a small boy: Another boy, Tshaguli-mbuangga, had been stealing Tshava's things from his house. Tshava complained of this to his own father. The father watched for Tshaguli-mbuangga and when he saw him enter the house he gave a signal to Tshava. Tshava then killed Tshaguli-mbuangga with a spear. His father helped him.

In Tshava's account of the matter there is no mention of any annoyance on the part of the murdered boy's relatives, but this no doubt occurred. In any case Tshava presented valuables to them. He states that he paid: One *Turbo* shell, one mother-of-pearl crescent, one tortoise-shell arm band, and one necklace of small *Conus* shells.

The relatives of the boy whom he killed presented a *tambointsha*, i.e. a tassel, to Tshava. This he will later exhibit on his lime stick (cf. Plate XXIVA) as a symbol of successful homicide. But as yet his initiatory grade have not been promoted to the wearing of this type of ornament.

2. While I was in the village of Komindimbit watching initiation ceremonies, a woman, a visitor from another village, was detected spying upon the flutes from a Malay Apple tree up which she had climbed. I heard a lot of shouting in the ceremonial house and dashed to the scene. I found a number

[1] Cf. Case 9, p. 69.

of men jumping about with sticks in their hands while others
remained quietly sitting on the platforms. The active angry
group were the members of the moiety B, especially Bx 4
and By 4 (cf. Diagram, p. 245). These were the groups which
had just paid for the privilege of seeing the flutes and therefore the offence was regarded as specifically against them,
not as against the whole group of men of that ceremonial
house. One man of moiety A did, it is true, give a many-
pronged spear to a member of moiety B, saying, "Go on,
spear her!" But this was the only part they took in the matter.
A party of the angry young men dashed off with sticks in
their hands to search the village for her. But they returned
without beating her—I think they could not find her. The
remainder of the group contented themselves with dancing
and jumping about in the ceremonial house with sticks and
spears.

Finally the members of moiety A quieted them and per-
suaded them to accept compensation. The two leaders or
nambu wail (literally "heads of crocodiles"—an initiatory
grade is called a "crocodile") of Ax 4 and By 4 accordingly
stood while bunches of areca nuts with valuables attached
were placed upon their shoulders. The valuables were: two
mother-of-pearl crescents and two *Turbo* shells. These would
later be used in buying a pig for some feast.

3. While I was in Kankanamun a scandal occurred in the
mbwole or junior ceremonial house. A man named Wi-ndjuat-
mali saw, in the middle of the night, a couple copulating
there. He identified them as T. and T.'s wife. It was none
of Wi-ndjuat-mali's business because he belongs to the group
B 4 and therefore is a member of the big ceremonial house
and has nothing to do with the *mbwole* which belongs especially
to the groups Ax 5 and Ay 5. But there is a great deal of
rivalry between the moieties A and B and it is possible that
Wi-ndjuat-mali, as a member of B, took a certain malicious
pleasure in telling the A group of the misfortune which had
overtaken their *mbwole*. In any case Wi-ndjaut-mali told a
member of A 5 what he had seen and the latter immediately
told the rest of A 5. T., the culprit, was himself a member
of this group and thus it came about that a row was precipitated *inside* the group A 5.

[1]T. was summoned by means of the slit-gong in the *mbwole* but he did not come. Nothing was done till the evening. Then the gong was beaten again and T. came. Then they debated against him: "Why did you copulate in our *mbwole*? You have made us 'cold'. One and all we are utterly bad; and never again shall we be all right. Another time if you stay in your house (not answering the gong) we shall go up into it and beat you there. We shall put spells upon your wife so that she runs away to another man."

Then T. said, "Make Wi-ndjuat-mali come quickly. They are joking; they are lying against me." But Wi-ndjuat-mali said, "It is no lie, it is true."

Then they said, "Your wife—we will kill a pig (and pay for it) with her property. We are children with no property. When the elder grade call for a pig, we will kill it with his property.

"She is not a woman who copulates with a small vulva. Her secretions have spilt in our *mbwole*. And she is a pregnant woman too. Yes. The semen that you have ejaculated will cover up the child. They will have to scrape the child when it is born."[2] And so on.

A deaf and dumb man also took part in this debate and expressed himself in noises and gestures with great force; but his remarks were untranslatable.

Of the various speakers against T. in this debate, all were members of moiety *A* with the exception noted in the text of Wi-ndjuat-mali (*B* 4) who spoke only as a witness. Most of the speakers were members of *A* 5, but one or two men who were half-promoted from *A* 5 into *A* 3 also spoke. The debate went on far into the night.

A few days later I asked some questions about the incident. To my question, "Why did T. and his wife use the *mbwole* when they had a house of their own?" I received the reply, "He and his wife had a long love affair before they were married. They used then to use all sorts of places and so they still go on doing the same thing."

---

[1] The following is a translation of a dictated account of this debate. Explanatory phrases not in the original have been inserted in brackets.

[2] Whenever this happens it is suggested that the parents failed to keep the taboo on copulation during pregnancy.

To my question, "Are they going to kill one of T.'s pigs or take any of his property?" the reply was "No. Of course not. *He is a member of their own initiatory group.*"

This last statement was the crux of the matter and the lack of any punitive mechanism was due simply to this fact. But such cases in which no sanction can be applied are comparatively rare. The situation was exceptional owing to the whole scandal being connected with the *mbwole* where only one of the two moieties *A* and *B* is represented and where also the moieties *x* and *y* are not sharply pitted against each other. Had a corresponding outrage been committed in the big ceremonial house the resulting row would probably have been between the moieties *A* and *B*, the various grades of each moiety taking sides against the opposite moiety. Other types of segmentation might, however, have occurred in the big ceremonial house. The building is subdivided among the various clans and the moieties *x* and *y*. Thus the offence, if it occurred in a particular part of the ceremonial house, might have been phrased as an offence against a particular clan, or as an offence of *x* against *y*.

4. While I was in Palimbai a quarrel broke out in a house which was shared by two classificatory brothers, Koulavwan and Membi-awan (cf. Genealogy). Both of these brothers were married, but the wife of Membi-awan had recently borne a child which she was suckling. Membi-awan was therefore unable to sleep with her. Koulavwan went away for one night to trade with a bush village leaving his wife, Kaindshi-mboli-agwi, behind in Palimbai. On his return, the following afternoon, he was told by his mother Kapma-tshat-tagwa and the other women of the house that Membi-awan had gone in the night to Kaindshi-mboli-agwi's mosquito bag.

It appears that what actually happened in the night was as follows: Membi-awan went to Kaindshi-mboli-agwi's bag, opened the end of it and put his hand in and there felt her head. She woke up and shouted and Kapma-tshat-tagwa came out of her bag to see what was the matter. Membi-awan then hid among the pots in the house; but he was identified.

This led immediately to three more or less separable quarrels:

(*a*) Between the two classificatory brothers Koulavwan and
Membi-awan. Koulavwan took his complaint to the village
"Tultul", a native appointed by the Government to act as
an interpreter to the District Officer. This official heard what
both parties had to say but did nothing about it. Membi-awan

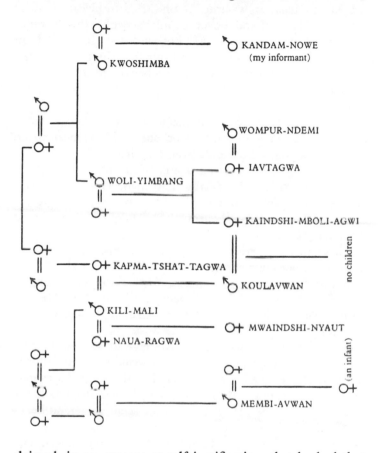

claimed, in an attempt at self-justification, that he had that
day seen Kaindshi-mboli-agwi give a betel leaf to her sister's
husband, Wompur-ndemi, and that his visit to her mosquito
bag was in order to verify that his brother's wife was being
faithful. This excuse was brushed aside. She said, "You
wanted to copulate with me. Now you are lying."

Koulavwan insulted Membi-awan, saying, "Yes, when we went to work for the white man, you were only a cook[1] but I *worked*."

(*b*) Between Koulavwan and his wife. He beat her "because she had not herself come and told him, but had waited till Kapma-tshat-tagwa told". She tried to placate him by saying, "My husband never copulates with other people; I am going to be like that. My husband is straight and I am the same." Further, she pointed out that she had expected something of the sort to happen and had asked the little girl, Mwaindshi-nyaut, to come and sleep with her in her bag for protection.

But my informant, her classificatory brother, said to Koulavwan, "Go on and beat my sister; she ought to have hurried up and told you." So Koulavwan beat her.

(*c*) This quarrel between Koulavwan and his wife had reverberations in another family: Kili-mali, sympathising with Koulavwan, started to beat his own wives and children, saying, "They were all asleep. They ought to have heard what was happening."

These three quarrels apparently died down before night and all the persons concerned went to bed as usual.

But, in the middle of the night, Koulavwan got up, angry, out of his mosquito bag and went and jumped on Membi-awan's bag in which the latter was sleeping. Then the two men fought.

After that, still dissatisfied, Koulavwan started to beat Kaindshi-mboli-agwi again. Both husband and wife fell into the water under the house—it was then the flooded season—and she screamed for Kandem-nowe who was sleeping in the next house but one. He dashed to the scene to defend her—"We are both of the same *nggwoil-nggu*."[2]

Then Woli-yimbang and Kwoshimba came on the scene and

---

[1] I.e. a cook for the labour line. The office of cook in a European's house is superior, but the cook for the labourers is despised.

[2] This term is applied to a close patrilineal group within the clan. The word is, I think, a plural of the word *nggwail* (father's father or son's son), an exception to the usual rule according to which the Iatmul language lacks plural suffixes. The only other similar words that I know of are: *nyan-nggu* (children) and *mbwambo-nggu* (mother's fathers).

began to take Kaindshi-mboli-agwi's part. They scolded Koulavwan—"Kaindshi-mboli-agwi is your mother. You ought
to hit her more gently." (This terminology is based on the
fact that Kaindshi-mboli-agwi is Woli-yimbang's daughter and
Woli-yimbang is a classificatory brother of Kapma-tshat-
tagwa. Therefore Kaindshi-mboli-agwi is Koulavwan's *wau*'s
daughter, a relative who is called *nyame*, mother.)

Woli-yimbang took a paddle, and Kwoshimba took the
stalk of a sago leaf. With these weapons they made a show of
beating Koulavwan.

Then Kili-mali, Koulavwan's father, came up and was angry
with Woli-yimbang and Kwoshimba. He said they had no
right to interfere. "Kaindshi-mboli-agwi is no longer your
daughter. She belongs to us."

In reply the two fathers of Kaindshi-mboli-agwi at once
started to complain about the quality of the valuables that
had been paid to their clan as a bride price for Kaindshi-
mboli-agwi; and Kili-mali replied with comments upon the
nggwat keranda (presentation of valuables from the bride's
people to those of the husband). This led to general re-
criminations on the subject of their mutual indebtedness.
Kwoshimba pointed out that he was always helping Kili-mali
with his *ngglambi* (i.e. paying out valuables so that Kili-mali
might make sacrifices to the *wagan*). "In fact", said Kwoshimba, "I am your elder brother."

Kili-mali denied all this indebtedness and said that Kwoshimba had always been a poor man till Koulavwan married
Kaindshi-mboli-agwi and gave him some mother-of-pearl
crescents.

Kwoshimba then referred to an incident in the distant
past, when he had half-killed a man and permitted Kili-mali
to complete the homicide and claim the achievement: "I
really killed that man, but you wore the homicidal paint for
him. How many have you killed for yourself?"

Finally Koulavwan intervened between Kili-mali and Kwoshimba, saying, "Our fathers must not begin fighting." The
disputants then ceased, and Kwoshimba and Woli-yimbang
said, "True, there is our blood in Koulavwan" (i.e. "Koulavwan is related to us through his mother"). Then they

exchanged areca nuts with Kili-mali as a sign of reconciliation; and all went back to bed.

These four examples of the Iatmul handling of offences, together with those which have been given in the chapter on "Sorcery and Vengeance", will suffice to show the principles and diffuse nature of the system. In the present context, that of enquiry into the sociological functions of affinal links, the conclusions which I draw from this material may be phrased as follows:

In every society, divergences from the cultural norm are liable to threaten the integration of that society, and this is clearly evident among the Iatmul. But under the Iatmul system the type of disintegration which is threatened is somewhat different from that which threatens our own societies.

In our own case, the threat is one of increasing muddle and confusion in our cultural norms, a rot which might spread through the community, producing first a lack of orientation among the individuals and finally the collapse of the society. Occasionally, in cases where the divergence has some positive quality bringing with it a new orientation, the divergent individuals unite together to form a group within the major community, a group with different cultural norms from the rest of the population. When this occurs, society can often accommodate the resulting group provided only that it is not too active in its opposition to the centres of the centripetal system. Alternatively, the divergent group may either succeed in altering the cultural system of the society or it may go off to some other part of the world and there found a culture *different* from that of the parent group.

But among the Iatmul the position is very different. Here there can be no threat of general confusion resulting from a blurring of cultural norms, since the chronic state of the culture is one in which the norms are weakly defined.[1] Neither can there be a conflict between an organising centre and divergent groups, because there is no organising centre in the purely peripheral system of the Iatmul. Rather the threat is one of *fission* of the community.

[1] The extremely drastic circumstances of modern culture contact with the white man have, however, had an effect somewhat of this nature. Cf. the story of Tshimbat, p. 168.

We have seen how every quarrel leads to the integration of loosely defined rival groups around the protagonists. It is this process which continually threatens the integration of the Iatmul village. With more profound rivalry the groups may grow until, in extreme cases, the fission actually takes place. One of the contending groups then leaves the village and founds a colony elsewhere, but with the *same* set of norms as the parent community.

A further difference between the two systems is implicit in what I have already said: in our own community the sub-groups which split from the parent are usually, perhaps always, formed of individuals whose divergent behaviour is based upon some form of *doctrine*. The danger to the *status quo* in our communities is the man who discards the cultural norms and who, either articulately or by example, "teacheth other men so". Among the Iatmul, who, it seems, normally regard the rules as things to break if you are strong enough, this situation does not arise. Their fissions spring not from conflicting doctrines but from rivalry between individuals or groups.

Such then is the danger which continually threatens the Iatmul community; and when this sociological position is stated the relevance of the stressing of the affinal links to the integration of the community is at once evident. These links form a network which runs *across* the patrilineal systems of clans, moieties, and initiatory groups and thereby ties the conflicting groups together. More than this, the patterning and ubiquity of the affinal relationships are such as to ensure that whenever a quarrel reaches serious dimensions there shall be some individuals marked out to act as peacemakers, to intervene between the disputants as Koulavwan in the example given above intervened between his father and his *wau*.

CHAPTER VIII

Problems and Methods of Approach

Problems. In our endeavour to relate the details of the *naven* to their setting, we examined first the ceremonies themselves, then the basic formulations upon which the *wau* relationship is built, and lastly the function which the ceremonies perform in integrating the Society. The structural and sociological analysis has revealed two important facts: (1) the cultural structure provides a setting with which the ceremonies are consistent; and (2) the ceremonies perform an important function in stressing certain classificatory relationships and thereby contribute to the integration of Iatmul society; crudely stated—if there were no *naven* ceremonies and no sociological analogues of *naven*, the villages would not be so big. We have then two sorts of factors, structural and sociological, which are certainly contributory to the *naven* complex.

But in the study of equilibria—and all cultural and social anthropology is a study of equilibria—we can never be sure that all the relevant factors have been mentioned. It is exceedingly dangerous to point to certain factors and to say that they constitute the whole cause of the effect which we want to explain. In the present context it would have been dangerous to point to the structural facts upon which the *naven* complex is built and to say that the sociological aspects of the matter were irrelevant, and equally dangerous, intolerantly preferring the sociological view, to deny the importance of the structure. In such cases it is safer to say that the known factors *contribute* to the effect in question, and to go on with the search for other types of factors which may also contribute their quota to the equilibrium. It may often be difficult to decide whether all the details of an effect can be ascribed to the known causes, but in the present instance there are many points which are not adequately explained by either the structural or the sociological factors.

In the first place, there are still features inside the *naven* plexus itself for which we have not accounted. We still do not know why the antics of the *wau* should be so exaggerated, nor why they should be comic. Why in stressing his identification with the boy's mother should he dress himself as a filthy and disgusting old hag? Why does he put on the very worst of feminine garments while the father's sister puts on the very best male ornaments that she can find?

Secondly, there is the group of questions raised the moment we push our enquiry farther along the branches in this network of cause and effect. Our theory of reasons for the development of *naven* has been built up inside a closed system in which the size of the village, for example, was assumed *ex hypothesi*. But a stable culture is itself a complete functional system and therefore we cannot assume certain facts as fixed in this way, but must still try to relate them back to the other features of the culture until we have a completely circular or reticulate exposition of causes and effects. We must therefore ask: Why should the villages be so large?

This question cannot be answered out of hand, in the present instance, by a reference to head-hunting warfare as the factor which puts a premium on large size of village and close integration of the community. So far as the Iatmul culture itself is concerned, I might have been tempted to regard warfare as the complete answer to the question, but from what Dr Fortune tells me of the Mundugumor people on the Yuat River, a tributary of the Sepik, warfare can be at best only a contributory factor, since the Mundugumor people combine the particular traits which we are tempted to regard as functionally incompatible. They live in small communities which are further subdivided into small family hamlets; the integration of the community is at so low a level that individuals—even own brothers—are intensely jealous and hostile among themselves; lastly, they are keen and successful head-hunters and cannibals.

But more significant than either of the above types of question are the problems of individual motivation. Even if we were studying a community in which "civic sense" was very highly developed in the individuals, it is very doubtful whether we should be justified in regarding the sociological functions of their behaviour as indications of the underlying motives. In any case the Iatmul are but little given to expressing the reasons for their behaviour in sociological terms.

Similarly, if we were studying a people who set great value upon the logical consistency of their behaviour we should hesitate before accepting their logical reasons for behaviour as honest statements of motive. We may well look askance at the father who tells us that he is beating his child either "for the good of the State" or "because it is the logical thing to do".

But though neither structural nor sociological phrasings can be accepted as statements of motive, one fact has been mentioned in passing which we might accept as a sufficient statement of the *wau*'s motives for performance of *naven*. We have seen that as a result of *naven* the *wau* gains the allegiance of his *laua* and we might dismiss the problem as solved in those terms.

Such a solution assumes, however, that the *wau* wants allegiance, and to me this assumption is unsatisfactory. Does he want allegiance because all male human beings are so born that they must later crave for the allegiance of their fellows? Or has the *wau* been so moulded by the culture in which he has lived that he wants allegiance?

If we discard the genetical view of the matter and suppose that the desire for allegiance is a conditioned appetite implanted in individuals by their culture, it then follows that this desire may take very various forms in different cultures. The possession of many loyal supporters may be associated with the most various forms of gratification. For example, an individual may want the allegiance of his fellows because he is conditioned to feel an agreeable emotional warmth when he and his friends are all jolly together. Alternatively, he may be so conditioned that, lacking loyal friends, he feels anxious and afraid of a spear in his back, while their presence gives him confidence. Or again, the possession of loyal friends may be a conditional stimulus for the reactions associated with the gratification of pride. The crude assumption that all men desire the allegiance of their fellows ignores all these various possibilities and so tells us virtually nothing about the *wau*'s motives.

At this point we may note that all the questions which have been raised in this chapter are closely inter-related. We have asked: (1) Why is the *wau* a buffoon? (2) Why are the villages large? (3) What are the *wau*'s motives? All these questions might be answered in terms of emotional satisfaction given to the individuals by the various phenomena, buffoonery, large villages and *naven*.

But we have noted in an earlier chapter (p. 32) that the study of affective functions is beset with difficulty and entails a preliminary study of the ethos of the culture; it will therefore be useful at this point to examine the ethological method in rather more detail and to consider the relationship between ethology and the concepts of philosophical history from which it is derived.

Zeitgeist and Configuration. The historical approach to the cultures of primitive peoples has been frequently abused by students who practise the methods of functional anthropology. It has been said that the historians are solely concerned with the search for origins and with the construction of speculative narrative. This abuse is only justified if the writings of the historians emphasise this aspect of their work at the expense of more scientific aspects. History, in so far as it is a science, is concerned not with narratives and origins but with *generalisations* from narrative, generalisations based upon the comparative study of the processes of cultural and social change. The major achievement of, for example, the heliolithic historians is, not their theory that almost all the cultures of the world have been derived from those of Egypt and Sumeria, but the picture which they have given us of the processes of change and degradation in culture. It is indeed high time that some student set to work upon the classification of these processes.

In the description of diachronic process, historians have used many of the same functional and economic concepts as are used in synchronic studies; but they have also elaborated one concept which has only very recently been adopted into the vocabulary of synchronic anthropology. This is the concept of *Zeitgeist*, the spirit of the times, a concept which owes its origin to the Dilthey-Spengler school of philosophical history.

The suggestion of this school is that the occurrence of cultural changes is in part controlled by some abstract property of the culture, which may vary from period to period so that at one time a given change is appropriate and occurs easily though a hundred years earlier the same innovation may have been rejected by the culture because it was in some way inappropriate.

Dr Benedict[1] has developed a related concept, that of the "configuration" of culture, and has done some very interesting and important work on primitive cultures. She has shown, for instance, that the refusal of the Zuni to adopt either peyote or alcoholic drinks was conditioned by the Apollonian configuration of their culture, while the neighbouring peoples with Dionysian cultures adopted both these stimulants with enthusiasm.

In European history the same type of concept is invoked in order to explain such curious facts as that Leonardo's mechanical inventions passed unnoticed or were laughed at during his life-time, and their importance was only gradually recognised during the following three centuries; or that the theory of evolution, though stated many times, did not obtain general acceptance until the industrial revolution had made the world "ready" to receive it.

In handling these concepts of *Zeitgeist*, configuration, etc., it is difficult to define their essential meaning without invoking some sort of mysticism. Their exponents have in general taken the wiser course, illustrating the concepts with concrete examples rather than giving abstract definitions of the terms which they use, and it is unfortunate that even this

---

[1] Benedict, "Psychological Types in the Cultures of the South-west", *Proc. 23rd Internat. Congress of Americanists*, 1928, pp. 572–581; *Patterns of Culture*, New York, 1934. I have been much indebted to the ideas of Dr Benedict and at the end of this book an attempt will be made to indicate the relations between my concepts, *ethos* and *eidos*, and her concept of configuration and the terms Apollonian or Dionysian, realist or non-realist, which she uses to describe cultural emphases.

course has led to their being branded as mystics. There are, however, certain generalisations which seem to apply to all these concepts. In the first place the concepts are in all cases based upon an holistic rather than upon a crudely analytic study of the culture. The thesis is that when a culture is considered as a whole certain emphases emerge built up from the juxtaposition of the diverse traits of which the culture is composed.

If we examine the content of these emphases we find that they are conceived to be either *systems of thought* or *scales of values*.

But the two words, *thought* and *value*, are terms which have been snatched from the jargon of individual psychology; we must therefore consider in what sense a culture may be supposed to possess either a system of thought or a scale of values. At the present time, we must follow the opinion of the majority of psychologists in dismissing the theory of the group mind as unnecessary, and therefore regard all the thinking and feeling which occurs in a culture as done by individuals. Thus when we attribute a system of thought or a scale of values to a culture, we must mean that the culture in some way affects the psychology of the individuals, causing whole groups of individuals to think and feel alike.

There are two ways in which culture might do this, either by education, inducing and promoting certain types of psychological process, or by selection, favouring those individuals who have an innate tendency to psychological processes of a certain kind. In the present state of our knowledge of genetics, we cannot pretend to estimate the relative importance of these two methods of changing the psychology of a population. We can only suppose that both the method of selection and the method of education are at work in every community. For convenience, I shall dodge the issue of choosing between the two hypotheses by using a non-committal term which shall subsume them both. Following Dr Benedict I shall speak of culture as *standardising* the psychology of the individuals. This indeed is probably one of the fundamental axioms of the holistic approach in all sciences: that the object studied—be it an animal, a plant or a community—is composed of units whose properties are in some way *standardised* by their position in the whole organisation. The time is not yet ripe for any detailed analysis of the possible standardising effects which culture may have upon the individuals in the community, but we may say at once that culture will affect their scale of values. It will affect the manner in which their instincts and emotions are organised into sentiments to respond differentially to the various stimuli of life; we may find, for example, that in one culture physical pain, hunger, poverty and asceticism are associated with a heightening of pride, while in another, pride is associated with the possession of property, and in another again, pride may be even gratified by public ridicule.

The effects of the culture upon the system of thought of the individuals are, however, not so clear. That the circumstances of a man's life will affect the *content* of his thought is plain enough, but the whole question of what we mean by a *system* of thought remains to be elucidated. I shall therefore leave this question for examination in a later chapter and proceed with the study of the standardisation of the affective aspects of the individual's psychology.

Psychological Theories and Ethology. With this theory, that a culture may standardise the affective make-up of individuals, we may now turn to the theories of those who have sought to explain social phenomena upon psychological grounds. These theories are based upon broad statements that human beings, men or women or both, in all races and all parts of the world, have certain fixed patterns of emotional reaction. In applying such a theory to our *naven* ceremonies we might for example say that men have *naturally* certain attitudes towards women and that therefore whenever men dress as women their behaviour is exaggerated into buffoonery; while women, on the other hand, are affected in their own special way when they dress up as men and therefore they put on a prodigious amount of swagger. Or again we might say that "human beings are naturally gregarious" and that this fact is a complete and sufficient explanation of the large size of the Iatmul village. When faced with the small size of Mundugumor villages we may say that their smallness is due to the natural hostility which exists between males.

When stated in this way the theories have a slightly ridiculous appearance, but it is worth while to consider the position in which we should find ourselves if we indulged the facile building of these theories to an unlimited extent. We should find that we had attributed to the human race a large number of conflicting tendencies and that we had invoked certain tendencies in the interpretation of one culture and other, perhaps opposite, tendencies in the interpretation of another. Such a position is untenable unless we have some criterion whereby we may justify our choice of a particular psychological potentiality for use in interpreting a particular culture—some criterion whereby we may decide which potentialities may legitimately be invoked in describing a given culture. But, inasmuch as human beings often appear to harbour conflicting tendencies and potentialities, this position, with all its contradictions, may become tenable as soon as a satisfactory criterion is discovered.

Such a criterion may, I believe, be derived from the conclusion at which we arrived above in our examination of the concepts of the historian. We concluded that culture standardises the emotional reactions of individuals, and modifies, the organisation of their sentiments; in fact, that culture modifies the very same aspects of the individual which are invoked by the rough and ready psychological theories of culture. We must therefore re-phrase the psychological theories in some such terms as these: A human being is born into the world with potentialities and tendencies which may be developed in very various directions, and it may well be that different individuals have different potentialities. The culture into which an individual is born stresses certain of his potentialities and suppresses others, and it acts selectively, favouring the individuals who are best endowed with the potentialities preferred in the culture and discriminating against those with alien tendencies. In this way the culture standardises the organisation of the emotions of individuals.

So long as we bear in mind this process of standardisation, we may safely invoke, in order to explain the culture, the sentiments of the individuals; but we must always verify that the sentiments invoked are actually those which are fostered by the culture in question. In the case of Mundugumor culture, if it can be shown that the hostility between individuals is an aspect of human nature which is as a matter of fact stressed by the culture, then we shall be justified in referring to this hostility as a factor which contributes to cause the people to live in small villages. In the same way, if it could be shown that Iatmul culture stressed man's gregarious tendencies, we should be justified in regarding the gregarious sentiment or instinct as important in the moulding of the culture. As a matter of fact, however, this facet of human nature is *not* specially stressed in Iatmul culture and this explanation must therefore be discarded.

I shall show later that pride is so stressed and that the pride is of a sort to be gratified by the big ceremonial houses which require organised labour on a large scale, by great ceremonies and dances which require many performers and by head-hunting which prospers when the village is strong in numbers. Thus the large size of the village serves an important function in gratifying pride—an attribute of human nature which is much stressed in Iatmul culture and to which therefore we are justified in referring.

The essence of the method is then that we first determine the system of sentiments which is normal to the culture and emphasised in its institutions; and when this system is identified we are justified in referring to it as a factor which has been active in shaping the institutions. It will be observed that the argument is circular.

In part the circularity is due to a characteristic of all scientific methods—the fact that we must observe a number of comparable phenomena before we can make any theoretical statement about any one phenomenon. But the circularity in the present case is also in part due to the nature of the phenomena which we are studying.

If we are studying jealousy and the institutions which regulate sexual life, we may argue both that the institutions stress jealousy and that jealousy has shaped the institutions. It would seem indeed that circularity is a universal property of functional systems, and that it may be recognised even in such crude and simple systems as the machines devised by man. In the motor-car for example the magneto produces electricity because the engine is running and the engine runs because of the sparks provided by the magneto. Each element in the functional system contributes to the activity of the others and each is dependent upon the activity of the others.

As long as we take an external—behaviouristic—view of a functional system we can avoid statements of circularity. We can see a motor-car as a thing into which petrol is poured and which runs along the road producing smoke and killing pedestrians. But the moment we turn from this external view and begin to study the internal workings of the functional system we are forced to accept the fundamental circularity of the phenomena. And this acceptance is demanded not only by ethology but by the whole functional approach to anthropology; and the students who are engaged in working from this point of view have realised this. Thus Malinowski claims that "the functional view avoids the error of attributing priority to one or the other aspect of culture. Material objects, social grouping, traditional and moral values, as well as knowledge are all welded into a functional system."[1]

A further and more compelling argument in favour of the circular or reticulate view of functional systems is to be found in the fact that any other view would drive us to belief either in a "first cause" or in some sort of teleology —in fact we should have to accept some fundamental dualism in nature which is philosophically inadmissible.[2]

Thus, since the phenomena which we are studying are themselves inter-dependent it is certain that our descriptions must contain inter-dependent statements; and since this is so the descriptions must for ever be regarded as "not proven" unless we can devise some method of transcending the limits of the circles. In a functional analysis we subdivide the systems studied into a series of parts or elements and produce theories about the functional relationships between these

---

[1] *Encyclopaedia Britannica*, art. Anthropology.
[2] Cf. Whitehead, *The Concept of Nature*, 1920, especially chap. II.

elements. So long as we study a single system these statements are bound to be circular and therefore not proven. But if we could extract comparable parts from different systems and verify that a given element has the same function in different systems we could finally verify the statements.

The orthodox functional school has adopted the practice of dividing cultures into *institutions*. But since the same institution is liable to have the most various functions in different societies the final verification of the theories is impossible. If we take the institution of marriage we find that it may function variously in the determination of status of offspring, in the regulation of sex life, in the education of offspring, in the regulation of economic life, etc.; and we find that the relative importance of these functions in different cultures varies so widely that it is almost impossible to verify by comparative methods the truth of any statement which we may make about marriage in any one culture.

The ethological approach involves a very different system of subdivision of culture. Its thesis is that we may abstract from a culture a certain systematic aspect called ethos which we may define as the expression of *a culturally standardised system of organisation of the instincts and emotions of the individuals*. The ethos of a given culture is as we shall see an abstraction from the whole mass of its institutions and formulations and it might therefore be expected that ethoses would be infinitely various from culture to culture—as various as the institutions themselves. Actually, however, it is possible that in this infinite variousness it is the *content* of affective life which alters from culture to culture, while the underlying systems or ethoses are continually repeating themselves. It seems likely—a more definite statement would be premature—that we may ultimately be able to classify the types of ethos.

The psychologists are already at work on the grading and classification of individuals and already it seems certain that different types of individual are prone to different systems of organisation of their emotions and instincts. If this be so, then there is a strong probability that the types of ethos will fall into the same categories of classification as the individuals and therefore we may expect to find similar ethos in different cultures and shall be able to verify our conclusions as to the functional effects of any one type of ethos by comparison of its expression in one culture with its expression in another, thus finally transcending the limits of the circular argument.[1]

EXAMPLES OF ETHOS IN ENGLISH CULTURE

Before describing the ethos of Iatmul culture, I shall illustrate the ethological approach by some examples taken from our own culture in order to give a clearer impression of what I mean by ethos. When a group of young intellectual English men or women are talking and joking together wittily and with a touch of light cynicism, there is established among them for the time being a definite tone of appropriate behaviour. Such specific tones of behaviour are in all cases indicative of an ethos. They are expressions of a standardised system of emotional attitudes. In this case the men have temporarily adopted a definite set of sentiments towards the rest of the world, a definite attitude towards reality, and they will joke about subjects which at another time they would treat with seriousness. If one of the men suddenly intrudes a sincere or realist remark it will be received with no enthusiasm—perhaps with a moment's silence and a slight feeling that the sincere person has committed a solecism.

---

[1] In a discussion of this matter which I have had with Professor F. C. Bartlett, he maintained that the position is not really circular but "spiral"; and that therefore this attempt both to justify and to escape from the circularity is irrelevant. But he is also sceptical of the validity of isolating synchronic from diachronic enquiries and therefore does not think in purely synchronic terms. The isolation is, I agree, artificial, but I think that it is useful and necessary in the present state of our science. Later, when we know more about the diachronic aspects of culture, it may be possible to synthesise the two methods of approach. It is probable that we shall then see that what appear to be circles in a synchronic "transverse section" of a culture are really spirals. Such a synthesis will of course let us out of the unproven circular argument. But if we are to escape from it *without* altering our synchronic point of view, this escape must be by way of the comparative method, the classification of the various ethoses and comparison of the ethoses with their associated cultural systems.

On another occasion the same group of persons may adopt[1] a different ethos; they may talk realistically and sincerely. Then if the blunderer makes a flippant joke it will fall flat and feel like a solecism.

The point which I wish to stress in this example is that any group of people may establish among themselves an ethos which as soon as it is established becomes a very real factor in determining their conduct. This ethos is expressed in the tone of their behaviour. I have deliberately for my initial example chosen an instance of labile and temporary ethos in order to show that the process of development of ethos, far from being mysterious and rare, is an everyday phenomenon. The same group of intellectuals were at one time serious and at another witty, and if the blunderer had had sufficient force of personality he could have swung the group from one ethos to the other. He could have influenced the evolution of ethos within the group.

But if, instead of such a temporary conversation group, we examine some more formed and permanent group—say an army mess or a college high table—whose members continually meet under the same conditions, we find the ethological position much more stable. In the more casual groups sometimes one sort of remark and sometimes another is inappropriate, but in any formed group we find certain types of remark, certain tones of conversation permanently taboo. The ethoses of the formed groups are still not absolutely fixed. The processes of ethological change are still at work and if we could compare a college high table or an officers' mess of fifty years ago with those groups as they are to-day we should no doubt find very considerable changes. Such changes are only very much slower in the formed groups, and enormously greater force of character or force of circumstances is required suddenly to shift the ethos.

Correlated with this greater stability of ethos, there is a new phenomenon present in the formed groups which was

---

[1] The factors—weather, mood, outside events, interactions of personalities, etc.—which cause these changes of ethos have never been investigated. Their study, though laborious and difficult, will certainly throw light upon many of our problems.

absent or scarcely recognisable in the unformed. The group has developed its own cultural structure and its own "traditions" which have grown up hand in hand with the ethos. At the high table (a group which is more familiar to me than is the army mess) we find such cultural developments as the Latin Grace, the dons' gowns and the silver presented to the College by former generations of Fellows. All these things have their effect in emphasising and stabilising the ethos of the group; and we cannot in any instance say that a given detail is due exclusively either to tradition or to the present ethos. The dons of St John's College drink water, beer, claret, sherry and port—but not cocktails; and in their choice they are guided both by tradition and by the ethos of the group. These two factors work together and we may say that the dons drink as they do both because generations of dons have drunk on the same sound system in the past and because actually in the present that system seems to them appropriate to the ethos of their society. Whatever detail of the tradition we examine, the same considerations apply. The Latin Grace, the architecture of the college, the snuff after dinner on Sundays, the loving cup, the rose water, the feasts—all these cultural details constitute an intricate series of channels which express and guide the ethos.[1] The details were in the past selected by the ethos and are still preserved by it. The system is a circular one; and the very attitude which the dons adopt towards the past has been historically formed and is an expression of their present ethos.

This intimate relationship between ethos and cultural structure is especially characteristic of small segregated groups where the ethos is uniform and the "tradition" very much alive. Indeed when we say that a tradition is "alive" what we mean is simply this, that it retains its connection with a persisting ethos. But when we come to consider not isolated groups but whole civilisations we must expect to find much more variety of ethos and more details of culture which have

---

[1] Such metaphors as this are of course dangerous. Their use encourages us to think of ethos and structure as different "things" instead of realising as we should that they are only different *aspects* of the same behaviour. I have let the metaphor stand *pour encourager les autres*.

been separated from the ethological contexts in which they were appropriate and retained as discrepant elements in an otherwise harmonious culture. Nevertheless I believe that the concept of ethos may valuably be applied ever to such enormous and confused cultures as those of Western Europe. In such cases we must never lose sight of the variations of ethos in different sections of the community and the curious dovetailing of the ethoses of the different sections into an harmonious whole, whereby, for example, peasants with one ethos are enabled to live happily under feudal lords who have a different ethos. Differentiated systems of this kind may persist for generations and only break down when the scales of values are questioned; when the lords begin to doubt the ethics of their position and the serfs to doubt the propriety of submission—phenomena which are liable to occur when the differentiation has proceeded too far.

CHAPTER IX

The Ethos of Iatmul Culture: the Men

COMPARED with European ethology, the conditions among the Iatmul are remarkably simple, since their culture recognises no differentiation of rank or class. Indeed the only social differentiation which we need to consider is that which exists between the sexes, and, since the problems which we are studying are connected with transvesticism, it is the differentiation between the sexes which is most likely to afford clues.

From whatever side we approach the culture, whatever institutions we study, we find the same sort of contrast between the life of the men and that of the women. Broadly, we may say that the men are occupied with the spectacular, dramatic, and violent activities which have their centre in the ceremonial house, while the women are occupied with the useful and necessary routines of food-getting, cooking, and rearing children—activities which centre around the dwelling house and the gardens. The contrast between ceremonial house and dwelling house is fundamental for the culture and will serve as the best starting-point for ethological description.

The ceremonial house is a splendid building, as much as a hundred and twenty feet in length, with towering gables at the ends (see Plate VII A). Inside the building, there is a long vista from end to end down the series of supporting posts as in the nave of a darkened church; and the resemblance to a church is carried further in the native attitudes towards the building. There is a series of taboos on any sort of desecration. The earth floor must not be scratched nor the woodwork damaged. A man should not walk right through the building and out at the other end; he should turn aside and pass out by one of the side entrances. To walk right through the building is felt to be an expression of overweening pride—as if a man should lay claim to the whole building as his personal property.

But the analogy between ceremonial house and church must not be pushed too far for many reasons: the ceremonial house serves not only as a place of ritual but also as a clubhouse where men meet and gossip and as an assembly-room where they debate and brawl. Further, the ceremonial house does not stand to the natives as a symbol of their devotion but rather as a symbol of their pride in head-hunting. Where we think of a church as sacred and cool, they think of a ceremonial house as " hot ", imbued with heat by the violence and killing which were necessary for its building and consecration. Lastly, the ethos of behaviour in the ceremonial house is as far removed from the austerity which we associate with certain churches as it is from the meek devotion associated with others.

Instead of austerity or meekness, there is a mixture of pride and histrionic self-consciousness.[1] An important man on entering the ceremonial house is conscious that the public eye is upon him and he responds to this stimulus by some sort of over-emphasis. He will enter with a gesture and call attention to his presence with some remark. Sometimes he will tend towards a harsh swagger and over-consciousness of pride, and sometimes he will respond with buffoonery. But in whichever direction he reacts, the reaction is theatrical and superficial. Either pride or clowning is accepted as respectable and normal behaviour.

In this community there are no steady and dignified chiefs —indeed no formulated chieftainship at all—but instead there is continual emphasis on self-assertion. A man achieves standing in the community by his achievements in war, by sorcery and esoteric knowledge, by shamanism, by wealth, by intrigue, and, to some extent, by age. But in addition to these factors he gains standing by playing up to the public eye; and the more standing he has, the more conspicuous

---

[1] In my description of ethos, I have not hesitated to invoke the concepts of emotion and to use terms which strictly should only be used by observers about their own introspections. I have been driven to this loose phrasing through lack of any proper technique for recording and of any language for describing human gesture and behaviour. But I wish it to be understood that statements of this kind are an attempt—crude and unscientific perhaps—to convey to the reader some impression of the *behaviour* of Iatmul natives.

will be his behaviour. The greatest and most influential men will resort freely either to harsh vituperation or to buffoonery when they are in the centre of the stage, and reserve their dignity for occasions when they are in the background.

Among the younger men whose standing is not as yet assured, there is rather more self-control. They will enter the ceremonial house soberly and unobtrusively and sit there quietly and gravely in the presence of their ranting seniors. But there is a smaller junior ceremonial house for the boys. Here they carry out in miniature the ceremonial of the senior house and here they imitate their elders in mixing pride with buffoonery.

We may summarise the ethos of the ceremonial house by describing the institution as a club—not a club in which the members are at their ease, but a club in which, though separated from their womenfolk, they are acutely conscious of being in public. This self-consciousness is present even at times when there is no specific formal or ritual activity taking place, but it is enormously more marked when the men are assembled in the ceremonial house for some debate or ritual performance.

Any matter of general interest may be disputed formally in a traditional manner. In every large ceremonial house there is a special stool, which differs from the ordinary stools on which men sit in having a "back", like a chair, carved into some representation of totemic ancestors. This stool is not used for sitting upon, and is indeed not casually touched if it be an old and sacred specimen.[1] It is used solely as a

[1] As far as I could make out the degree of sacredness of the debating stool depends simply upon its age. I was told that the stool acquires "heat" through the rage of successive debaters.

In the account which I have published of Iatmul debating (*Oceania*, 1932, p. 260) a slip occurs. I there state "on the occasion of quiet careful discussion of personal names and genealogies *Dracaena* leaves are used, but in more passionate disputes, bunches of coconut leaflets are substituted. These latter being tougher are more able to withstand violent usage." This information was given me in reply to a question as to the reasons for using *Dracaena* leaves. I had only seen debates in which coconut leaves were used, but when visiting the ceremonial house in another village I found on the debating stool three bunches of *Dracaena* leaves and therefore asked the question. Actually, inasmuch as my later observations of debates about names and ancestors do not bear out the statement that such debates are "quiet", I now incline to the view that

table for debates. The speaker has three bunches of *Dracaena* leaves, or coconut leaflets. He picks these up at the beginning of his speech and with the combined bunches he gives a blow to the stool. He then puts down the bunches on the stool, one by one, as if they were a tally of his sentences. When all are put down, he again bunches them together and gives another single blow. This series of actions is repeated throughout his speech, ending with a final blow.

The tone of the debates is noisy, angry and, above all, ironical. The speakers work themselves up to a high pitch of superficial excitement, all the time tempering their violence with histrionic gesture and alternating in their tone between harshness and buffoonery. The style of the oratory varies a good deal from speaker to speaker and that of the more admired performers may tend towards the display of erudition or towards violence or to a mixture of these attitudes. On the one hand there are men who carry in their heads between ten and twenty thousand polysyllabic names, men whose erudition in the totemic system is a matter of pride to the whole village; and on the other hand there are speakers who rely for effect upon gesture and tone rather than upon the matter of their discourse. Such a man will make a speech in which there is only the barest minimum of contribution to the issue—and that minimum something which has already been said by other speakers—but he will fill out his speech with assertions of his scorn and threats that he will rape the members of the opposition, accompanying the word with obscene pantomimic dance. Meanwhile the insulted will watch and smile a little, or laugh aloud and shout ironic encouragement to the insulting speaker. Besides these two types—the erudite and the abusive—there are also nervous and apologetic speakers whose contributions to the debate are despised. These men generally attempt the erudite style, but their memories are undermined by nervousness and their "howlers" are laughed at by the audience.

As the debate proceeds, both sides become more excited and some of the men leap to their feet, dancing with their

the reason given for the use of *Dracaena* leaves was an incorrect rationalisation and suspect that these leaves are used in debates about ancestors because of the ceremonial associations of the *Dracaena*.

spears in their hands and threatening an immediate resort to violence; but after a while they subside and the debate goes on. This dancing may occur three or four times in a single debate without any actual brawling, and then suddenly some exasperated speaker will go to the "root" of the matter and declaim some esoteric secret about the totemic ancestors of the other side, miming one of their cherished myths in a contemptuous dance. Before his pantomime has finished a brawl will have started which may lead to serious injuries and be followed by a long feud of killings by sorcery.

The emotions which are so dramatically paraded in debate have their centre in pride, and especially in individualistic pride. But hand in hand with this, there is developed a prodigious pride in the totemic ancestors of the clan; and most of the debates are concerned with the details of the totemic system. This totemic system has an obvious affective function—a very important one in this culture—of providing the members of every clan with matter for self-congratulation. But inversely the proud ethos of the culture has reacted in a curious way upon the system and though we are not here concerned with the origins of Iatmul totemism, a description of the system is relevant as indicating the emotional background against which we are to see the *naven* ceremonies.

The totemic system is enormously elaborated into a series of personal names, so that every individual bears names of totemic ancestors—spirits, birds, stars, animals, pots, adzes, etc. etc.—of his or her clan, and one individual may have thirty or more such names. Every clan has hundreds of these polysyllabic ancestral names which refer in their etymology to secret myths. It seems that the effect of pride upon this system has been to corrupt the origin myths, so that to-day each moiety has its own phrasing of the origins of the world according to which that moiety's own importance is stressed at the expense of the other moiety. The same tendency extends to the clans. Groups of clans flatter themselves by secretly maintaining that they are not really members of either moiety but are the *fons et origo* from which both moieties sprang; and each group has its own secret mythology to support its secret claims. The debates about totemism are usually concerned with attempts to steal totemic ancestors by stealing names and one of the most important features of every Iatmul ceremony (except initiation) is the chanting of name songs, whereby the members of the clan are reminded of the importance of their ancestors and the system is continually memorised.

Actually, as a result of the overlapping mythology and the stealing of names, the system is in a terribly muddled state. In spite of this, the people are very proud not only of the number of their totemic ancestors and their esoteric exploits in the origins of the world but even of the "straightness" of their song cycles. They feel that the whole gigantic system is perfectly schematic and coherent. Thus the prevailing pride which has led them to build up this mass of fraudulent heraldry is still such that the people regard the resulting tangle as rigid and coherent.

If we turn to the ritual connected with the ceremonial house, we see the men, as a group, still vying with each other, but in spite of their rivalry managing to work together to produce a spectacle which the women shall admire and marvel at. Almost without exception the ceremonies of the men are of this nature; and the ceremonial house serves as a Green Room for the preparation of the show. The men put on their masks and their ornaments in its privacy and thence sally forth to dance and perform before the women who are assembled on the banks at the sides of the dancing ground. Even such purely male affairs as initiation are so staged that parts of the ceremony are visible to the women who form an audience and who can hear issuing from the ceremonial house the mysterious and beautiful sounds made by the various secret musical instruments—flutes, gongs, bull-roarers, etc. Inside, behind screens or in the upper storey of the ceremonial house, the men who are producing these sounds are exceedingly conscious of that unseen audience of women. They think of the women as admiring their music, and if they make a technical blunder in the performance, it is the laughter of the women that they fear.[1]

---

[1] For a sketch of the technique and sociological setting of flute music in Iatmul culture, cf. *The Eagle*, St John's College Magazine, 1935, vol. XLVIII, pp. 158–170.

The same emphasis on pride occurs in many other contexts of the life of the men. Here it must be remembered that the ceremonial house is also the meeting place in which a great deal of the everyday work of the men's lives is organised. In it their hunting, fishing, building and canoe cutting are discussed; and these activities are carried out in the same spectacular manner as the ritual. The men form parties in their big canoes to go fishing or hunting; or they go off together in big groups to cut trees in the bush. Such parties are called together by rhythms beaten on the great slit-gongs in the ceremonial house; and if the work is taking place close to the ceremonial house the gongs are beaten to "put life"[1] into the workers. Finally, the completion of every considerable task is marked by the performance of some spectacular dance or ceremony.

Thus it comes about that the ritual significance of the ceremonies is almost completely ignored and the whole emphasis is laid on the function of the ceremony as a means of celebrating some labour accomplished and stressing the greatness of the clan ancestors. A ceremony nominally connected with fertility and prosperity was celebrated when a new floor had been put into the ceremonial house. On this occasion the majority of informants said that the ceremony was being performed "because of the new floor". Only a very few men were conscious of, or interested in, the ritual significance of the ceremony; and even these few were interested not in the magical effects of the ceremony but rather in its esoteric totemic origins—matters of great importance to clans whose pride is based largely upon details of their totemic ancestry. So the whole culture is moulded by the continual emphasis upon the spectacular, and by the pride of the male ethos. Each man of spirit struts and shouts, playacting to convince himself and others of the reality of a prestige which in this culture receives but little formal recognition.

---

[1] The native phrase, which is here translated "put life", is *yivut taka-*. *yivut* is the ordinary word for "movement" or "liveliness" and may be used as noun, adjective or verb; e.g. *vavi yivut yi-rega-nda*, literally "bird move going-it", i.e. "the bird is moving"; *yivut kami*, "a live fish". *taka-* is the ordinary word for "to put", "to set", etc. The constant references to *yivut* in contexts of work, play and ceremonial should be mentioned as characteristic of the *tempo* (cf. p. 255) of the culture.

No account of life in the ceremonial house would be complete without some reference to the ethos of initiation. Here we might expect in another culture to find the men combining together with dignity and austerity to instruct the youths; and in the painful process of scarification we might expect to find them inculcating Spartan resistance to pain. The culture has many of the elements which would seem appropriate to such an ascetic ethos: there are days during which neither novices nor initiators may eat or drink; and there are occasions on which the novice is made to drink filthy water. Again the culture contains elements which would make it appear that the novice is passing through a period of spiritual danger. He must not touch his food with his hand and he is subjected to a drastic washing which suggests ritual purification; and so we might reasonably expect to see the initiators protecting the novices from dangerous contamination.

But actually the spirit in which the ceremonies are carried out is neither that of asceticism nor that of carefulness; it is the spirit of irresponsible bullying and swagger. In the process of scarification nobody cares how the little boys bear their pain. If they scream, some of the initiators go and hammer on the gongs to drown the sound. The father of the little boy will perhaps stand by and watch the process, occasionally saying in a conventional way "That's enough! that's enough!" but no attention is paid to him. The operators are chiefly interested in their craft and regard the wriggling and resistance of the novice as prejudicial to it. The spectators are rather silent with, I think, a touch of "cold feet" at the sight of the infliction of pain divorced from the normal setting of histrionic excitement. A few are amused.

When pain is inflicted in other parts of initiation, it is done by men who enjoy doing it and who carry out their business in a cynical, practical-joking spirit. The drinking of filthy water is a great joke and the wretched novices are tricked into drinking plenty of it. On another occasion their mouths are opened with a piece of crocodile bone and examined "to see that they have not eaten what they ought not". They are not under any food taboos at this time, but the result of the examination is invariably the discovery that the mouth is unclean; and the bone is suddenly jabbed against the boy's gums making them bleed. Then the process is repeated for the other jaw. In the ritual washing, the partly healed backs of the novices are scrubbed, and they are splashed and splashed with icy water till they are whimpering with cold and misery. The emphasis is upon making them miserable rather than clean.

In the first week of their seclusion, the novices are subjected to a great variety of cruel and harsh tricks of this kind and for every trick there is some ritual pretext. And it is still more significant of the ethos of the culture that the bullying of the novices is used as a context in which the different groups of the initiators can make pride points against each other. One moiety of the initiators decided that the novices had been bullied as much as they could stand and were for omitting one of the ritual episodes. The other moiety then began to brag that the lenient ones were afraid of the fine fashion in which *they* would carry out the bullying; and the lenient party hardened their hearts and performed the episode with some extra savagery.

The little boy's introduction into the life of the ceremonial house is conducted on such lines as these and it is one which fits him admirably for the irresponsible histrionic pride and buffoonery which is characteristic of that institution. As in other cultures a boy is disciplined so that he may be able to wield authority, so on the Sepik he is subjected to irresponsible bullying and ignominy so that he becomes what we should describe as an over-compensating, harsh man—whom the natives describe as a "hot" man.

The natives themselves have summed up the ethos of initiation in a formulation which is especially interesting in view of the contrasting ethos of the two sexes. During the early period of initiation when the novices are being mercilessly bullied and hazed, they are spoken of as the "wives" of the initiators, whose penes they are made to handle. Here it seems that the linguistic usage indicates an ethological analogy between the relationship of man and wife and that of initiator and novice. Actually resort to such sadistic treatment only occurs in rather extreme circumstances in the case of wives, but it is perhaps true that the men would like to believe that they treat their wives as they do their novices. I think we may see a consistent cultural pattern running through the contrasting sex ethos, the shaming of the novices, the *wau* shaming himself by acting as the wife of the *laua*, and the use of the exclamation "*Lan men to!*" (husband thou indeed!), to express contemptible submission. Each of these elements of culture is based upon the basic assumption that the passive role in sex is shameful.[1]

In fact, the initiatory situation is not a simple one from an ethological point of view but is essentially a contact between two ethoses, that of the initiators and that of the novices. The ethos of the former is clearly a mere exaggeration of that of men in the daily life of the ceremonial house. But the ethos of the novices is not so clear. To some extent, especially in the early stages of initiation, they play the part of women; and we may ascribe some of the exaggeration of the initiators' behaviour to the presence in the ceremonial house of novices with an opposite ethos.

For the moment we are not concerned with how this contrast has arisen; but we may suppose that the little boys have to some extent imbibed the women's ethos in their early life and so come to their initiation with some of the emotional attitudes characteristic of women in this culture. This assumption I cannot definitely state to be founded on fact, because I did not study the children; but whether it is correct or not it is certain that there is some vague idea of this sort behind the initiation rituals. The response of the initiators to this real or nominal contrast between the novices and themselves is to force the boys further into the complementary position, dubbing them "wives" and bullying them into expressions of the wifely role.

The end of all this is the adoption by the novices of the masculine ethos, but it seems that the first step in inducing this process is to compel the novices to behave as women, a sufficiently paradoxical method of setting about the business to force us to enquire more closely into the processes which it involves.

---

[1] For an instance of the opposite opinion, cf. footnote, p. 142.

I believe that we may distinguish[1] four contributory processes:

1. A process whereby the novice becomes contra-suggestible to the female ethos. We have seen that the treatment of novices *qua* wives is a great deal more drastic and arbitrary than the treatment of real wives; and further that while the women grow up gradually into their ethos, the novices are suddenly and violently plunged into submission. These differences account I believe for the fact that, while the wives accept a somewhat submissive ethos without too much difficulty, the novices become contra-suggestible and rebel against it. It is certain that revolts and refusals to undergo further bullying sometimes occur among the novices. Such revolts may from our present point of view be regarded as symptoms of well developed contra-suggestibility to the submissive role, and the sharp repressive measures to which the initiators have recourse do not, we may suppose, extinguish the resentment in the breasts of the novices.[2]

2. A process whereby the novices become proud of the male ethos. We may suppose that the novices derive some feeling of superiority from their separation from mothers and sisters and especially from undergoing an experience from which women and smaller children are excluded. Certainly they begin at quite an early stage, even before their cuts have healed, to be proud of the scars. About ten days after the initial scarring the novices are taken out into the bush and are threatened with bamboo knives. When I saw this done, four out of the five novices shrank away and screamed with something approaching hysterics at the idea of further cutting and were excused, but the fifth novice after a few moments' hesitation submitted to the operators and proudly acquired extra scars without flinching.

---

[1] The theory here presented to a considerable extent overweighs the facts upon which it is based. This analysis of the initiatory process is, however, intended as a sample of the ethological point of view and an indication of the sort of problems which this point of view raises, not as an exposition of proven hypothesis.

[2] Nowadays it is often necessary to initiate boys who have been away to work for the white man and who return as grown men still unscarred. Under these circumstances such revolts are especially common and I was twice called upon by the initiators to help in the preservation of the system.

After the first week of intensive bullying, the relationship between initiators and novices alters. The former are no longer described as "elder brothers" or "husbands" but are now known as "mothers" of the novices. Corresponding to this change of phrasing, the initiators turn their attention to making much of the novices. They hunt game so that their "children" may grow big with good feeding, they teach them how to play the flutes and they undertake as communal tasks the making of various ornaments—pubic tassels, lime boxes, spear-throwers, etc. for presentation to the novices. Finally at the end of initiation the novice, decked in all this finery, is exhibited to the women as the hero of the occasion and the completion of his initiation is celebrated with *naven*.

3. Reactions to the presence of later novices. The ceremonial end of his initiation by no means completes the assimilation of the novice into the group of the initiators. There are other younger boys who though too young for initiation are by descent of the same grade as the recently initiated novices (cf. Fig. 4, p. 245). Later when these boys come up for initiation the former novices will line up with them to be shown the flutes all over again. They will not undergo again the more drastic forms of bullying but ceremonially they will be reckoned as novices. At all these ceremonies they are frankly bored. Only much later when they become themselves initiators will they begin to take considerable interest in the business of initiation. Then they react to the presence of novices by themselves becoming bullies, completely assimilated into the system.

4. Reactions to the presence of other initiators. The drastic bullying behaviour of the initiators is not merely a reaction to the presence of novices, but is also promoted to a great extent by the feelings of emulation which exist between members of the rival initiatory groups. While group Ax 3 (cf. Fig. 4, p. 245) is initiating By 4, Ay 3 is initiating Bx 4 on the opposite side of the ceremonial house and between these two initiating groups there is constant bickering and rivalry in the manner of their performance. On one occasion the novices had been so severely bullied that one of the bullying tricks, called *tshimangka*, the fish, in which the

novices would be very severely slapped, was postponed from day to day. Finally the time drew near when the novices should undergo ritual washing after which the bullying would cease. There were still several incidents in the programme of bullying to be carried out and the principal spokesman of group *Ax* 3 suggested that the *tshimangka* incident should be omitted. The immediate retort of group *Ay* 3 was a boasting taunt to those who suggested leniency. They suggested that *Ax* 3 were afraid of the brutal way in which they (*Ay* 3) would carry out the rite. In the face of this accusation *Ax* 3 hardened their hearts. The *tshimangka* was immediately undertaken and carried out. When it was over and while the novices were spitting the blood out of their mouths, the formerly lenient spokesman assured me that the rite had been performed with extra violence in reply to the gibes of *Ay* 3.[1]

One characteristic of the Iatmul men is very clearly exhibited in the contexts provided by the initiatory system. This is their tendency to "cut off their own noses to spite the other fellow's face". Whenever a serious situation arises in the initiatory enclosure, e.g. when a woman sees something of the secrets, or some disrespect is shown to the secret objects, or when some serious quarrel breaks out, the talk is always of "breaking the screens", throwing the whole initiatory system open to men, women and children, showing everything to everybody. Much of this talk is of course mere shouting without any real intention of drastic action, but from time to time an impasse causes such a degree of exasperation and shame that the men carry out some self-humiliating act which may cripple the ceremonial life of the village for some years.

Such a case occurred in Mindimbit. The initiatory system in this village was getting feebler and feebler. The boys were going away to work on plantations, leaving the village too weak in numbers for any great ritual to be attempted; while those who returned to the village did so with a scorn of the

[1] It is possible that a fifth factor—rivalry between villages—also contributes to shape the behaviour of the initiators. The men of Komindimbit boasted to me that their initiatory crocodile had more fierceness (*kau*) than that of any other village on the river; and it was, I think, more savage than those of Palimbai and Kankanamun.

initiatory crocodile. The ceremonial houses were full of boys
with no scars on their backs and one small ceremonial house
had even been slowly invaded by women and was now
deserted by the men, who left it to the women as a place
where they might sit and gossip.

One day, a party of Mindimbit natives were being given
a lift on a white recruiter's schooner. In the basket of one
of the young men was a small bamboo end-flute.[1] This
object was noticed by the man's wife, who picked it up and
said "What is this?" An older man was present and saw
the incident. He scolded the woman and she was ashamed.
Then he went to the ceremonial house and reported what
had happened and abused and ranted against the carelessness
of the young man who had allowed the accident to occur.
The old man and the other members of the opposite moiety
raided the young man's house and smashed his wife's pots.
But still they were not satisfied, so they went to the *tagail*
or junior ceremonial house[2] and to the dwelling houses, to
collect all the small boys together. Even the small toddlers
who could only just talk were included. Then, that night,
they showed them everything, including the *wagan* gongs.
The procedure was an extreme telescoping of the whole
initiatory cycle, but was unaccompanied by any sort of
scarification. This drastic event was regarded both in Mindimbit and in the scornful neighbouring villages as the final
utter shame and destruction of Iesinduma, the village "crocodile" of Mindimbit.

Another more violent but similar event took place in
Palimbai some fifty years ago. The village was celebrating
its *wagan*, a performance on the gongs very much more

---

[1] These end-flutes are toys made by work boys and are not indigenous
to Iatmul culture. But the men have decided that all wind instruments
ought not to be seen by women lest they should guess at the nature of
the flute music which they hear coming from the ceremonial house.

[2] The *tagail* differs from the *mbwole* in being the ceremonial house for
uninitiated members of $B\,4$ (cf. Diagram, p. 245), while the *mbwole*
belongs to $A\,5$. In most villages only one of these buildings is erected
and serves indiscriminately for both purposes. In Mindimbit there was
a new *tagail* which had been built in the hope that its presence might
attract the little uninitiated boys away from the senior ceremonial house.
But in this it failed. The *tagail* in Mindimbit was generally empty.

secret and serious than the earlier initiation ceremonies which only involve flutes, bullroarers, etc. In the *wagan* ceremonies, the secret gongs are beaten continuously,[1] day and night, for months on end, in the upper storey of the ceremonial house; and during all this time there must be no noise in the village: no person may quarrel or shout or break firewood. The spears stand, ever ready, leaning against the outside of the ceremonial house and against the screens, to kill any man or woman who offends against the *wagan* by disturbing the peace. But in spite of this preparedness for killing the spears are not often used, and the incident which occurred in Palimbai is still remembered vividly and is cited as proof that the spears are there for serious use.

The ceremony had proceeded without serious trouble and it was time to prepare for the final spectacle in which old men, impersonating the *wagan*, perform a dance in front of the assembled women. The initiatory group who were producing the spectacle went out to get croton (*Codiaeum*) leaves with which to ornament the giant representations of *wagan* (cf. Plate XXVIIIA), sneaking secretly out of the village so that no women should know of the methods of staging the spectacle. They had collected the leaves and put them in string bags and were on their way back to the village. Some children were playing near the mouth of the waterway which leads in from the river to Palimbai village, shooting the straight stalks of elephant grass with toy spear-throwers. One of these missiles fell in the men's canoe and pierced the string bag in which the croton leaves had been put.

The men at once gave chase and speared the small boy who was guilty of this offence. When they got back to the village a general fight ensued in which three (or four?) men were killed, all of them members of the clan Wainggwonda, "fathers" of the small boy.

---

[1] This is a statement of what *ought* to happen. Actually the rhythms are rarely maintained for more than two or three days and are generally interrupted by some quarrelling between the initiatory moieties. But the rhythm is started again as soon as each quarrel is over and so continues off and on for several months. When one performer is weary another takes his place, taking the moving gong stick from his hand so that not a beat is missed.

Then the clan, Tshimail, which had taken a chief part in the killing, went into the ceremonial house and pulled down the *wagan* gongs from the upper storey. They built a small screen around them in the dancing ground and then showed the gongs to all the women of Wainggwonda and handed over the sacred gong sticks to them, to keep in their houses.

In killing the boy, they had only acted according to the conventions of the *wagan* ceremony; but still, perhaps feeling they had gone too far, they were impelled to humiliate themselves, to preserve their pride. They did not hang *tambointsha* (tassels) on their lime sticks for these killings.

In the business of head-hunting, the masculine ethos no doubt reached its most complete expression; and though at the present time the ethos of head-hunting cannot be satisfactorily observed, there is enough left of the old system to give the investigator some impression of what that system implied. Lacking observations of actual behaviour my description must, however, be based on native accounts.

The emphasis here was not on courage; no better *coup* was scored for a kill which had entailed special hardship or bravery. It was as good to kill a woman as a man, and as good to kill by stealth as in open fight. An example will serve to illustrate this set of attitudes: In a raid on one of the neighbouring bush villages a woman was killed and her daughter was taken by the killer (Malikindjin) and brought back to Kankanamun. He took her to his house, where for a while he hid her, thinking of adopting her into his household. But she did not remain there. He took her to the ceremonial house and a discussion arose as to her fate. She pleaded that she should be pitied: "You are not my enemies; you should pity me; later I will marry in this village."

One of the young men, Avuran-mali, son of her captor, cut into this discussion, and in a friendly way invited her to come down to the gardens to get some sugar-cane. Accordingly he and the girl went down to the gardens together with one or two of the younger boys, among them my informant, Tshava, who was then a small boy. On arriving there, Avuran-mali speared her. (The duty of cleaning the skull fell to Tshava. An enemy skull must never be touched, and

Tshava had some difficulty in detaching a ligament. He therefore discarded the tongs, seized the end of the ligament in his teeth and pulled at it. His father saw him and was very shocked; but Tshava said to me: "The silly old man! How was I to know?"—an attitude towards taboos which is not uncommon among the Iatmul.)

But in spite of the lack of "sportsmanship", the activity of head-hunting was to a considerable extent a "sport". There was no clear rule that you must have a grudge against a man before you killed him, nor against a village before raiding it; though a majority of the killings were certainly regarded as vengeance. In general the fighting and killing was confined to the killing of foreigners, i.e. members of other villages, especially of villages against whom a feud existed. But even this rule was not too strictly interpreted; a woman, married into the village, might for purposes of head-hunting be considered a foreigner. I even came across one case in which a man wore a tassel for killing his own wife in revenge for a kill accomplished by members of the village from which she had come.

Two main motives informed this system, the personal pride of the individual and his pride and satisfaction in the prosperity and strength of his community. These two motives were closely tangled together. On the purely personal side, the successful homicide was entitled to special ornaments and paints and to the wearing of a flying fox skin as a pubic apron; while the apron of stripped *Dracaena* leaves was the reproach of the man who had never killed. The homicide was the hero of the most elaborate *naven* and the proud giver of feasts to his *lanoa nampa* (husband people). Lastly, he was admired by the women; and even to-day the women occasionally make scornful remarks about the calico loin-cloths worn by the young men who should strictly still be wearing *Dracaena* aprons like those which were given them when they were little boys being initiated.

The association of personal pride with success in head-hunting and of shame with failure is brought out too in the behaviour of those whose relatives had been killed. Their first duty was the taking of *nggambwa* (vengeance). The rings of cane worn in mourning for the killed individual may not be put aside until vengeance has been achieved; and a pointed reference to an unavenged relative is one of the most dangerous insults that one Iatmul can use in ranting against another—an insult which is felt to be specially aggravating now that head-hunting is forbidden.

Indeed so serious is the condition of those who are unable to secure revenge, that it produces *ngglambi* in the group and may lead to the sickness and death of its members.

This spreading of clan dysphoria resulting from the unavenged insult to the pride of the clan may be contrasted with the "sociological" phrasings of the benefits which successful head-hunting confers upon the community. Here, as is usual in sociological phrasings, the matter is expressed in tangled symbolism, but may be made clear by an artificial separation of the various components of the system:

1. The enemy body was, if possible, brought back to the village and was there ritually killed by a man wearing a mask which represents an eagle. Thus the kill symbolically became the achievement not only of the individual homicide but of the whole village.

2. The natives articulately say that the eagle is the *kau* of the village. *Kau* is a word which means "a raiding party", "a fighting force", "an expression of anger", etc. The eagle is also represented on the finial of the ceremonial house (cf. *Oceania*, 1932, Plate VIII), and at the ceremony with which this eagle is put in place, the bird speaks. He looks out over the enemy country and sees them there as "birds preening themselves" or as "fish jumping in the water"—ready to be killed.

3. The natives say that prosperity—plenty of children, health, dances and fine ceremonial houses—follows upon successful head-hunting.

4. Prosperity is also dependent upon the *mbwan*, those ancestral[1] spirits which are represented by standing stones.

[1] The *mbwan* are regarded as ancestors and are classified roughly with the *angk-au* or potsherd spirits. But in some cases at least, the *mbwan* are really the spirits not of deceased ancestors but of killed enemies. Perhaps they are thought of as ancestors because of their activity in promoting the proliferation of the community.

5. The heads of the killed were placed upon the *mbwan* and in some cases their bodies were buried under the *mbwan*.

6. The standing stones are phallic symbols, e.g. in the shaman's jargon the phrase for copulation is *mbwan tou-*, "setting up a standing stone".

7. The male sexual act is definitely associated with violence and pride.

Running through this plexus of cultural details we can clearly see the general position of head-hunting as the main source of the pride of the village, while associated with the pride is prosperity, fertility and the male sexual act; while on the opposite side of the picture but still a part of the same ethos, we can see the association of shame, mourning and *ngglambi*.

Closely linked with these emphases upon pride and shame is the development of the spectacular side of head-hunting. Every victory was celebrated by great dances and ceremonial which involved the whole village. The killer was the hero of these and he was at the same time the host at the feasts which accompany them. Even the vanquished assented to the beauty of the dances, as appears from a text collected in Mindimbit describing the typical series of events on a raid:

(After the fighting) they leave off. Then he (the killer, standing in his canoe and holding up the head which he has taken from the enemy) asks "I am going to my beautiful[1] dances, to my beautiful ceremonies. Call his name." (The vanquished reply) "It is so-and-so that you have speared." (Or the victor will say) "This one is a woman" and they (the vanquished) will call her name (and they will cry to the victors) "Go. Go to your beautiful dances, to your beautiful ceremonies."

[1] In this text the phrase which I have translated as "beautiful dances, beautiful ceremonies" is of considerable interest. The native word for "beautiful" is *yigen*, a common Iatmul word which is used to describe an admired face or spectacle. The same word also occurs in the adverb *yigen-mbwa*, "gently", the opposite of *nemwan-pa*, "violently" (literally, "greatly"). The whole phrase is *yigen vi, yigen mbwanggo*, a poetical form built up out of the common everyday phrase *vi mbwanggo*, "a (triumphant) war dance". In this phrase, *vi* is the word for a particular sort of spear with many points, used in warfare, and *mbwanggo* is the ordinary word for any dance or ceremony. In the traditional diction this phrase is divided into two parallel phrases, a common trick of Iatmul poetic genius. (Cf. also *yigen kundi*, "quiet singing", p. 156.)

CHAPTER X

The Ethos of Iatmul Culture: the Women

IN the everyday life of the women there is no such emphasis on pride and spectacular appearance.[1] The greater part of their time is spent on the necessary economic labours connected with the dwelling house—food-getting, cooking and attention to babies—and these activities are not carried out publicly and in big groups, but privately and quietly. In the very early morning before dawn the women go out in their tiny canoes to tend the fish traps in which they catch the prawns, eels and small fish which form the staple supply of protein food. Each canoe is just big enough to carry a woman and perhaps her small child, and on the stern is a little fire in an old pot for the woman to warm herself by in the chilly dawn; for the business of examining the traps involves her wading about in water up to her breast and she will be cold when the job is done. The little fleet of canoes,

---

[1] There is some local difference in ethos between the Eastern Iatmul (Mindimbit, Tambunum, etc.) and the Central Iatmul (Palimbai, Kankanamun, etc.). Among the Eastern people the women wear large quantities of shell ornaments in their daily life, only removing them for such tasks as the tending of fish traps. These women have also a slightly prouder bearing than the women of Palimbai and Kankanamun, who normally wear very little ornament. This difference has probably some bearing upon the culture as a whole, and it is worth mentioning that among the Eastern group *iai* marriage, in which the initiative rests with the woman, is commoner than in Palimbai; and that it was in Mindimbit that I was shown a very interesting lime gourd. It is the custom there for men to scratch on their gourds tallies of their successful love affairs; but the gourd in question besides the ordinary tally had, incised upon it, a large representation of a vulva ornamented with geometrical designs. I asked whether it was carved there as an emblem of the mother moiety, but the owner replied with pride: "No, that refers to a woman. I did not want her, but she came to my mosquito bag and took the active role in sex."

In general there is the same sort of contrast between the ethos of the two sexes in both areas, but this contrast is most marked in Kankanamun, where the women are definitely shabby. They are a little smarter in Palimbai and markedly smarter in Mindimbit. The men of Palimbai are conscious of the difference between their women and those of Kankanamun, and attribute it to the better supply of fish which they get because their village stands on the banks of a lake.

each with its column of smoke rising in the half light, is a very pretty sight; but as they draw away from the village each canoe separates from the others as each woman goes to the part of the river where her traps are set. In this work there is none of the excitement which the men introduce into their fishing expeditions. Each woman goes off by herself to do her day's work. When she has tended the fish traps she will go and collect, for firewood, old dead stems of the elephant grass which lines the banks of the river. Then she will return to the village where she will attend to the cooking and the jobs of the house.

A single house is divided between two or three men related by patrilineal ties; and this division of the house is felt by the men to be very real, almost a matter for stiffness and formality. The man who owns one end of the house will avoid intruding upon his brother's or his son's residence at the other end[1] although there is no screen or wall dividing the house, only the big sleeping bags in the centre of the floor. But although typically the women of a house are not mutually related, they seem to be much less conscious of the divisions and will constantly bandy remarks the whole length of the house. Each woman has her separate cooking place with its fire basins set up close to the wall and the different wives of one man carry out their cooking independently, but still there is more ease in their mutual relations and less self-consciousness than is the case among the men.

The women's life is regulated by a three-day week, the middle day of each three being a market day.[2] The supply of fish and prawns is such that a quantity can be set aside, either kept alive in baskets immersed in the water or smoked. On a market day the women gather up these supplies and go off in their canoes to the bush villages where they barter

[1] It is my impression that a son will visit his father much more freely and casually than the father will visit the son; and it is likely that there is some trace of the same tendency between brothers, the younger visiting the elder more freely.

[2] This three-day week is characteristic of Palimbai and Kankanamun, but is not adhered to by the Eastern Iatmul whose villages are farther from those of their bush neighbours. The Eastern Iatmul hold their markets irregularly, on specially arranged days, the two parties meeting in the grass country which separates their villages.

the fish for sago. The market may be held actually in the bush village, but more often the women of the bush tribes come half-way along the road and the parties meet at some agreed spot. The chaffering is done easily and with a good deal of jolliness. The deals are small and there is very little haggling but a great deal of talk, not only about the matter in hand but about the events of the last few days in the various villages. At these markets men may be present, but so far as I know their presence has no quelling effect on the general ease of the women. I have, however, only attended markets at which men were present and so cannot state definitely what effect their presence has upon the women. The women's markets contrast sharply with the behaviour of the men when they are engaged in their more serious negotiations. In buying a sleeping bag or a canoe, each party tries to outdo the other in a pose of critical taciturnity, and in the majority of cases no business will be done.

Compared with the proud men, the women are unostentatious. They are jolly and readily co-operative while the men are so obsessed with points of pride that co-operation is rendered difficult. But it must not be supposed that the women are mere submissive mice. A woman should know her own mind and be prepared to assert herself, even to take the initiative in love affairs. In *iai* marriage (cf. pp. 88, 89) it is nominally the woman who makes the advances and who, of her own accord and uninvited, goes to the house of her *ianan*. It is said in Tambunum that in such cases the chosen man has no right to refuse such a proposal.

The same pattern is often followed in less formalised marriages in which the woman has no such nominal right. A typical case will illustrate the extent of the woman's initiative: I had gone with my native servants to look at some ceremonies in the neighbouring village of Aibom, a village of foreigners who are not regarded as true Iatmul but whose social system is very closely related to that of the Iatmul. On the day after my return from this expedition, a girl from Aibom arrived alone in Kankanamun. She enquired for members of her own clan and went to the house of one of my informants who was her clan brother. She told

him that she was in love with one of my cook-boys. Her clan
brother asked "Which cook-boy?" She explained that she
did not know his name. So they went together to a spot
where they could observe my domestic staff and the girl
pointed out the cook-boy whom she loved and he was thereby
identified. In the negotiations which followed the boy and
girl modestly avoided each other (so the cook-boy told me)
but her clan brother acted as intermediary and she lodged
in his house. The boy was definitely flattered by the proposal
and decided to accept it. He sent a series of small presents
to the girl, which she accepted. Very soon messages began
to come from Aibom demanding the bride price; and the
cook-boy was not rich. There were delays, and after about
a fortnight the girl returned to Aibom. The boy then put on
some show of anger and demanded from her relatives some
return for the presents which he had given her. But this
return was not forthcoming.

The point which I wish to stress in this incident is the
extraordinary courage which the girl showed in coming alone
to a foreign village, and the clarity with which she knew her
own mind. Her conduct was regarded as culturally normal
by the Iatmul.

As a further documentation of the respect which is paid
to women of strong and courageous personality, we may cite
here a traditional myth which was told to me in Mindimbit
in explanation of the head-hunting alliance between that
village and Palimbai. Both of these villages have a traditional feud with the village of Kararau which lies between
them.

Kararau were killing us. They speared women who went out
to get tips of wild sugar-cane, and women who went to get water-
weed (for pigs' food), and women who went to their fish traps.
And they shot a man, Au-vitkai-mali. His wife was Tshanggi-mbo
and (his sister was) Au-vitkai-mangka. They shot him and beat
the gongs (in triumph). Au-vitkai-mangka was away; she was on
the lake (fishing). Au-vitkai-mali went to his garden and they speared
him, and the sound of his gongs came (over the lake). She asked,
"Whom have they speared?" and (the people) said, "They have
speared your husband."[1]

[1] The confusion which occurs here between the two women is typical
of texts dictated by the Iatmul. It is evident that my informant was

Then she filled up a string bag with shell valuables and she (went to the ceremonial house and) said, "Men of this village, I have brought (valuables) for you." But they said, "No. We do not want them", and they were ashamed (because they had not dared to accept the valuables which she had offered as payment for assistance).

Then she went down into her canoe; she loaded the valuables into the canoe; she took off her skirt and put it in the canoe. Au-vitkai-mangka was in the stern and Tshanggi-mbo in the bows. The bag of valuables was in the middle of the canoe. She went up the river to Palimbai, because she had heard his gongs. The two of them went by night.

They sat leaning against the ceremonial mound (a place of refuge) in Palimbai, and they put the bag of valuables on the ground close to the mound. At dawn (the people of Palimbai) got up and saw (them). They were sitting stripped of their skirts, with their skirts on their shoulders.[1]

*The men of Palimbai said, "They are women of Kararau"; and they were for spearing them. The women said, "Why will you spear us?". Kaulievi (of Palimbai) saw and said, "Don't spear them"; and he said "Come". Then he beat the gong to summon all the men of Palimbai, Kankanamun, Malinggai, and Jentschan. The men of the four villages came together, and the women told them to debate. The men said, "What women are you?"; and the women said, "We are women of Ienmali." (Ienmali is the name of the old site of Mindimbit.)

Kaulievi said, "Tell your story"; and Au-vitkai-mangka said, "That is (the sound of) my brother's gongs coming"; and the men said, "Who speared him?".

Au-vitkai-mangka then (calling the names of the totems of the four villages) appealed to Kankanamun: "You! Crocodile! Wanimali!"; and to Malingai: "You! Crocodile! Kavok!"; and to Palimbai: "You! Pig! Palimbai-awan!"; and to Jentschan: "You! Pig! Djimbut-nggowi!". And she said, "I shall take away my bag of valuables."

She set out the valuables in a line; and the four villages accepted them. That night they debated, "Already tomorrow we shall raid them." Each of the four villages (brought) a fleet of canoes. They formed into one fleet on the Sepik River.*[2]

They (the men) gave a spear to Au-vitkai-mangka and the men of

describing the man's sister as hearing the sound of the gongs while she was fishing, but when she asks who has been killed the reply is "Your husband".

[1] The nakedness of the women in this context seemed so natural to me when I was told the myth, that I did not enquire into the reasons for it. I have no doubt, however, that this nakedness is the mark of the suppliant and that it is, in some degree, analogous to the nakedness of the women in *naven* when they lie down before the hero.

[2] The passage between asterisks has been condensed.

Palimbai gave another spear to Tshanggi-mbo. They gave one canoe to the two, a swift canoe; and the two women were in the centre of the fleet.

When they drifted down to the Kararau (reaches of the) Sepik, (the canoes took up formation[1]). The two women hid in the centre. Then the men shot an eel. It said "War". (A favourable omen; and here my informant reproduced the grunting of the eel.)

The two women came out (of the fleet). They (went forward and) sang dirges in midstream.[2] They were smeared with clay (for mourning) and the people of Kararau came out to spear them. But (the women) were going down stream in a swift canoe. They went for the setting of Palimbai's battle (i.e. the women acted as decoys). The Palimbai people killed the people of Kararau and they caught two men (alive) in their hands. Au-vitkai-mangka speared one of them. Tshanggi-mbo speared the other. They speared them all, every one of them.

They all went upstream together to Palimbai and there they beat the gongs. The two women beat the gongs.[3] In the morning the men cut the women's hair and oiled them and presented valuables to them. Then they brought the women in a fleet of canoes to Ienmali, and left them. Kaulievi said "Kararau are our enemies" and he came and set up a stone (in Ienmali, now removed to Mindimbit). And so Palimbai took Kararau for enemies. That is the stone, here, and the name of the stone is Kaulievi, an ancestor of Kepmaindsha. That is why Kepmaindsha came here, and Tonggalus too. (Kepmaindsha and Tonggalus were two men who had left their own villages as a result of quarrels.) Later when the two women died, they made a song about them; the clan Mwailambu (to which Au-vitkai-mangka and my informant belonged) made the song.

In the household, too, a woman may have considerable power and authority. She it is who feeds the pigs and catches the fish; and it is upon these activities that her husband chiefly depends for the wealth which helps him to make a splash in the ceremonial house. When a man is haggling silently over a canoe or sleeping bag, he will withdraw before concluding the deal, in order to consult his wife. And, judging by the things which the wife is reported to say in

---

[1] This phrase is substituted for a list of technical terms which do not concern us here.

[2] It is usual for widows to sing dirges for their husbands as they go in their canoes to and from their work (cf. p. 157).

[3] The beating of the gongs, like the business of head-hunting, is normally only done by men. The oiling, hair-cutting and presentation of valuables are expressions of respect.

such circumstances, it appears that wives hold the purse strings very tight. But the stubbornness of an absent wife makes a very convenient tool in the business of striking a bargain and I doubt whether the wives are really as "strong" as their husbands report.

In a few households, however, it is definitely the wife who "wears the trousers", and in two such cases the sympathy of outsiders went to the wife rather than to the henpecked husband. It was the wife's misfortune to have married a weakling.

But, as against the occasional instances in which women take up an assertive role and even participate in warfare, the more habitual emphasis of the women's ethos is upon quiet co-operative attitudes. Though the woman may take the initiative in sexual advances, it is the activity of the male which is stressed in the native remarks about copulation, while the part played by the female is despised. In the Iatmul language the ordinary verb for copulation and the jocular synonyms which are used for it are, so far as I know, all of them transitive and in their active forms refer to the behaviour of the male. The same verbs may be used of the female role, but always in the passive.[1]

Thus, in our study of the women's ethos, we find a double emphasis. For the most part, the women exhibit a system of emotional attitudes which contrasts sharply with that of the men. While the latter behave almost consistently as though life were a splendid theatrical performance—almost a melodrama—with themselves in the centre of the stage, the women behave most of the time as though life were a cheerful co-operative routine in which the occupations of food-getting and child-rearing are enlivened by the dramatic and exciting activities of the men. But this jolly, co-operative attitude is not consistently adopted in all contexts, and we have seen that women occasionally adopt something approaching the male ethos and that they are admired for so doing.

In the ceremonial activities of the women, the same double

---

[1] These verbs may also be used in the dual in the active voice. The pidgin English idom, "play", has not been adopted in literal translation into Iatmul.

emphasis is present, and these activities fall into two distinct ethological groups according as one or the other emphasis is predominant. In general, the jolly, co-operative emphasis is most evident when women celebrate by themselves in the absence of men, while the proud ethos is exhibited when women celebrate publicly in the dancing ground of the village with men in the audience.

In the first group are the frequent dances held by women in the houses. These dances are very much resented by the men, who regard them with contempt and do all they can to discourage them. When the women's ceremonial demands that the performers shall keep taboos on sexual intercourse, the men do their best to cause them to break these taboos— and then boast to the anthropologist of the postponement of the ceremony. At such times the sex opposition—never far from the surface—comes to a head. Quarrels between husbands and wives are especially frequent and the wives take their revenge by refusing to cook sago for their husbands.

I found a husband sitting sulkily in the ceremonial house. He had a lump of sago which he was roasting rather ineffectually, naked on the fire—for the men believe that the art of cooking pancakes of sago is one which they cannot learn. He said, "Yes, we copulate with them, but they never retaliate", a reference to the despised passive sexual role. Then he jumped to his feet and shouted this taunt across the village to the women in his house, from which he was excluded.

For, in spite of their contemptuous attitude, the men withdraw quietly enough when the dance is actually about to take place; and the women are left in complete command of the house. They remove all the sleeping bags, clearing the floor-space for the dance; and a great crowd of women collects together from all over the village, all joking and in the best of humour. After a while the dancing begins and the gathering sounds very gay.

On one occasion I was sitting in the ceremonial house when these sounds reached us. The men greeted the sound with contempt; but I asked if I could go and look at the women's dance. The men told me that the dances were very silly and not worth my looking at, that they could not compare with the dances of the men. I said that Mindimbit was a poor, " cold " village; the men never performed any of their vaunted spectacles; and if I couldn't see any dancing I was going to go to another village. Finally and very reluctantly a young man said he would take me to the dance and we went together. We entered while the women were dancing round the floor in short jumping steps and singing a rather catchy quick tune. We sat down on stools in a corner of the floor in silence. My companion was acutely uncomfortable and after a few minutes he slipped away.

When the women had finished the song they came crowding up to me and offered me areca nut and betel. I asked for lime. Most of them had never seen me chew betel and my acceptance of the offer created some excitement, screams of laughter and noisy screaming talk—like a flock of parrots. In the middle of this excitement two women started to dance in front of me. They stood face to face in a jumping dance and at every jump one woman pushed her hands forward with palms pressed together, the other woman received the hands between her own. Between the beats the first woman drew her hands back only to push them forward again at the next beat. This dance was obviously a representation of copulatory action, but I was completely surprised when suddenly, in a single jump, the two women dropped to a sitting position on the floor, still facing each other and one sitting between the legs of the other, one of the standard positions for sexual intercourse. In this position the women went through the motions of copulation still in time to the beat of the song; and then, as suddenly as they had sat down, they jumped up in a single motion and after a few more jumps they broke off, giggling.

The mere description of what these two women did gives very little idea of the extraordinary *naïveté* of this "obscenity" and the contrast between it and the harsher obscenity of the men. Lacking a photographic record I can only record my subjective impression of this.

After more joking the crowd of women left me to continue

their dancing round the house. The same jolly atmosphere continued and I had no doubt that this was the regular tone of the women's dances in the absence of men. I think too that an analysis of the tunes, sung by women and men respectively, would show the same ethological contrast between the sexes which I observed in their behaviour.

But this characteristic jollity is not carried over to those occasions when women celebrate publicly. Then they march in procession in the middle of the dancing ground before a mixed audience of men and other women, they are fully decorated and wear among other things many ornaments usually worn by men—a sort of mild transvesticism which will be referred to again in the theoretical analysis of *naven*. They march with a fine proud bearing very different from their jolly behaviour when men are absent, and different too from their quieter demeanour on everyday occasions when they are in the presence of men but are not decked out in finery. Their marching gait in these processions is indeed more closely comparable with their swaggering demeanour when dressed in full homicidal war paint for the *naven* ceremonies than with their patterns of behaviour on other occasions.

Examples of this proud bearing and of the mild transvesticism are shown on Plate XIX, and a similar phenomenon to that which appears on ritual occasions may be observed fairly constantly under the experimental conditions produced by pointing a camera at an individual. When a woman is photographed, her response to the camera depends on whether she is wearing her finery or is in everyday dress. In her finery she holds her head high when the eye of the camera is on her; but in everyday dress she hangs her head and rather shrinks from the public appearance constituted by standing up alone before the photographer while her friends are watching in the background (cf. Plate XXVI). When a man is photographed, whatever his costume, he tends to swagger before the camera and his hand goes almost instinctively to his lime stick as if about to make with it the loud grating sound which is used to express anger and pride.

CHAPTER XI

Attitudes towards Death

WE have so far examined the behaviour of men and women only in the everyday and the ceremonial contexts of their culture. But the contrast between the sexes is even more striking when the individuals are faced with events highly charged with emotion. To illustrate this I shall describe the sequences of events after a death has occurred.

In Palimbai I was woken up at about 4.30 a.m. one morning by the sound of weeping in the house next to mine. I went to see what was happening and found that a young man, who had been sick for two or three months, had finally died. The corpse was stretched out straight and was naked. A circle of women were crouched around it, and the mother of the dead man had the head in her lap. A fire was burning close to the corpse, and gave the only light in the house. The women were quietly weeping and dirging, singing songs of the dead man's maternal clan—songs which might be used on gay and everyday occasions, only now the singing was slow and out of tune and broken with sobs.

From time to time there were pauses when all were quiet, and then one of the women would make some remark about the dead man. Some incident of his life was referred to, or some small possession of his mentioned with the suggestion that it should be buried with him. Then the songs and the sobs were resumed, set off by this recall of another facet of their personal loss.

There was one man in the house. He was sitting apart from the group of women, silent and embarrassed. When I went up and spoke to him he greeted my intrusion with pleasure and was very ready to discuss the arrangements for the funeral—how the Government had forbidden them to expose the corpse in a canoe till the floods abated; as it was, they would have to take the body into the Tshuosh country to find dry ground for the burial; they would measure the body in order to know how big a hole to make; and so on.

The women's weeping continued until after dawn, but it was no hysterical exaggeration of grief such as is recorded from other primitive communities. My feeling was that I was witnessing an easy and natural expression of sorrow at a personal loss.

The behaviour of the man was in marked contrast to this. On the one hand, he quite evidently wanted to escape from his embarrassment into conversation about the funeral, and, on the other hand, he boasted: "We (Iatmul) are not people who only play at weeping", and when I asked whether it was not only the women who wept he felt this remark as an aspersion on the men, and insisted that men also weep. Later in the conversation he turned his attention from the affairs connected with the death and started to lecture me on the East wind and its totemic position.

After dawn, we first waited for the rain to stop and then proceeded with the funeral, for the men say: "Tears are not found in the lake", meaning that the supply will not last long and therefore they must bury the body quickly. The men put the body in a canoe and took it over the fens from one piece of supposedly higher ground to another, but all were flooded. We were a party of eight of whom two were women, the mother and sister of the dead man. The mother sat immediately behind the corpse, sometimes dirging over it. Conversation on the journey was quiet and concerned with possible causes of the death. Our plan was to take the body to Marap village, but the men were impatient and on the way said: "No, Marap is a long way. Let us bury him on Movat Tevwi" (a piece of high ground in the fens): but in the end we had to go to Marap and arrived there in the late afternoon. The Tshuosh were not pleased to see us, but at last they permitted the body to be buried under a deserted house.

The men had some difficulty in digging a grave and the site had to be changed twice because they came on other old bones when they dug. Finally the corpse was laid in the grave and the portrait skull of the dead man's brother was deposited with him. A shilling was placed in each of his hands, and his string bag was placed in the grave. Since he was buried among the Tshuosh, the grave was oriented so that (on raising his head) the corpse would look towards the setting sun as is the custom of the Tshuosh. In Palimbai the dead are normally buried with their feet towards the dancing ground, so that the corpse is not "looking into the bush".

The women retained their skirts throughout this burial, but I was told by an informant in Kankanamun that the mother, sister and wife of a dead man would normally be naked while he was being buried. This nakedness is, no doubt, in some way analogous to the nakedness of the women when they lie down before the hero in *naven*, and with the nakedness of female suppliants.

The mother of the dead man stayed behind in Marap to mourn for a few days, but the rest of us returned that evening to Palimbai as a normally cheerful party, no longer an embarrassed and silent group.

A second occasion on which I was able to observe the reactions of the men to a death was on the day after Tepmanagwan, a great fighter of Palimbai, had died. He died during the night and was buried in the early morning. I arrived in the village at about 9 o'clock, after the interment, and found that the men had by then left the grave and gone to the ceremonial house. A few women were weeping at the graveside, and from the ceremonial house I could just hear the weeping of others in the house of the dead man.

I proposed going to the house, but the men hinted that I should not. They were just starting a debate in the ceremonial house. It was a scandal that Tepmanagwan had died without passing on his esoteric knowledge, and the debate was an enquiry as to whose fault this was. A few men sobbed while they were making their speeches, and I found it hard to judge whether these sobs were the result of genuine feeling bursting its way to the surface against resistance, or whether they were a theatrical performance staged in absence of strong feeling to give this impression. In any case, it was perfectly clear that the men's sobs were very far removed from the natural weeping of the women.

The debate reached no conclusion and, when it petered out, the men set up a figure to represent the dead man (cf. Plate XX A). The head of the figure was an unripe coconut and the body was made of bundles of palm leaves. Spears were set up against the figure with their points stuck into it to mark where the man had been wounded in war, and other spears were stuck into the ground beside the figure for those which he had dodged. A series of vertical spears was set up in front of the figure according to his achievements. The figure itself was ornamented with shells, etc. Six sago baskets were suspended from its right shoulder to represent his six wives, a string bag was suspended on the left shoulder representing his skill in magic. A number of sprigs of ginger in its headdress represented persons whom he had invited to the village so that other people could kill them. In the right hand of the figure was a dry lump of sago, because it was said that in his lifetime he had once killed a bird by throwing a lump of sago at it. A branch of *timbut* (lemon) set in the ground beside the figure was symbolic of his knowledge of mythology. Finally, on the ground at the feet of the figure were a broom and a pair of boards used for picking up rubbish. These objects were symbolic of the work which the dead man had done in cleaning the ceremonial house during his lifetime.

This figure was set up by members of the initiatory moiety of which the deceased was a member. It was a boast of the greatness of their moiety, and when the figure was completed all the men of both moieties crowded round it. The members of the opposite moiety came forward one by one to claim equivalent feats. One man said: "I have a wound here on my hip, where the (people of) Kararau speared me. I take that spear", and took the spear set against the figure's hip. Another said: "I killed so-and-so. I take that spear", and so on till all the emblems of prowess had been removed.

Thus the men made, out of the context of death only a few hours old, an occasion for expressing the competitive pride of the initiatory moieties. They escaped entirely from a situation which was embarrassing because it seemed to demand a sincere expression of personal loss, an expression which their pride could scarcely brook. From this situation they took refuge in a cultural stunt. They re-phrased their attitude towards the dead and expressed it satisfactorily in terms of spectacular pride, the emotional language in which they are at ease.[1] But such a handling of grief is, I think, still not adequate, and later a further compensation is added. It is my impression[2] that when a man is asked about some past funeral he will generally drag into his answer some reference to his own great weeping, in spite of the fact that, at the time, he wept but little and probably made a show of his resistance to womanly tears.

In the later mortuary ceremonies the contrast between the behaviour of the two sexes continues. The skull of the dead man is exhumed and a portrait is modelled upon it in clay. This is set up one night as the head of a highly decorated doll which represents the deceased (cf. Plate XXI b). Around this figure the men stage an elaborate performance of name songs and flute music. The ceremony which is called *min-tshanggu* (cf. p. 47) takes place in a dwelling house and the women are present as audience. The flutes are played by men hidden under the platform on which the figure stands, whither they have been secretly smuggled. Thus though the ceremony takes place in a dwelling house and its context is a personal one, it is staged upon the same general principles as all the other performances staged by the men, a spectacle for the admiration and mystification of the women.

Later the women have a little mourning ceremony by themselves in the absence of men. This is called *yigen kundi* ("quiet singing"). It takes place at night in a house from which the men have withdrawn. A little food is hung up for the ghost to "eat" and the women sit in a circle by the firelight

[1] In our own culture, of course, both these types of emotional pattern and many others are mixed and tangled together in our mortuary ceremonial. Culture contacts and the recurring instability of Western European societies have provided us with every sort of conflicting phrasing, and these phrasings have been preserved for us in script through the ages. But the Iatmul have a less confused culture.

[2] Unfortunately I took no notes of such effusions, accepting them as a matter of course.

and softly sing the name songs of the dead man's mother's clan. The wife or mother of the dead may weep a little, but the general tone of the group is one of quiet sorrow rather than of passionate grief. The "quiet singing" goes on till late in the night, when the women disperse to their houses.

In the months or years which follow, the mother or wife of the dead man will occasionally, when she is alone, sing as a dirge one of the name songs of his maternal clan; these dirges may often be heard on the river, coming from some woman mourning as she paddles her canoe to her garden or fish traps. The men quite frequently caricature this musical effort, probably because the attitude of the women towards death is one which they themselves find distasteful.

Indeed one of the most important phenomena which is brought to light by examination of ethological contrast is this distaste which persons trained in one ethos, their emotional reactions standardised in one pattern, feel for other possible ethoses.[1] In the illustration which I gave, I mentioned how a remark which is out of tune with the temporary ethos of a group of Englishmen is received with silence, and in Iatmul culture we may recognise the same phenomenon in the distaste which the men feel for the ethos of the women. This phenomenon is extraordinarily widespread and it affects even the anthropologist whose task it is to be an impartial student of ethos. Every adjective which he uses is coloured by and evokes the feelings which one sort of personality has about another. I have described the ethos of the men as histrionic, dramatising, over-compensating, etc., but these words are only a description of the men's behaviour as seen by me, with my personality moulded to a European pattern. My comments are in no sense absolute statements. The men themselves would no doubt describe their own behaviour as "natural"; while they would probably describe that of the woman as "sentimental".

It is difficult too to describe a pair of contrasting ethoses

---

[1] The same phenomenon is to be observed also in the reactions of individuals trained in one system of cultural structure to other possible systems. They are, I think, liable to regard the other systems as non-sensical, illogical and perhaps tedious.

without so weighting the descriptions that one or the other appears preferable or more "natural". The business of the scientist is to describe relationships between phenomena, and any ethos which he finds in a culture must be regarded not as "natural" but as normal to the culture. Unfortunately what is normal to one culture may well be abnormal to another, and the anthropologist has at his disposal only the adjectives and phrases of his own culture. Thus it has happened that English people with whom I have discussed Iatmul ethos have sometimes remarked that the women appear to be "well adjusted" while the men appear to be "strained" and "psychopathic". My friends forget that the values assigned by European psychiatrists to various mental conditions are either *cultural* values based upon European ethos or estimates of the fitness of the individual for life in a European community.

The pride of the men, when seen in contrast with the women's ethos, may appear to my readers somewhat angular and uncomfortable. I found it also splendid. I have not stressed this aspect enough, and therefore I shall conclude the description of Iatmul ethos with a free translation of a story which illustrates how a man should behave when his own death stares him in the face:

A man went with his dog to hunt for wild pigs in the sago swamps. When they had killed a pig the man went to wash its guts in a lake. While he was doing this a giant crocodile (Mandangku, an ancestor of Tshingkawi clan) seized him by the instep and held him fast.

The man said to the dog, "Go home and sniff at my feather headdress, and sniff at my armbands and all my ornaments."

Then the dog went home and when the man's wife saw the dog sniffing at the ornaments she took them and put them in a basket, and the dog led her and her child back to where the man was, still held fast by the crocodile on the lake side.

When he saw them the man said, "My child, my wife, I am lost", and then he said, "Give me my things."

He put on his legbands and his shell girdle. He put on one of his armbands and then he put on the other. He hung his mother-of-pearl crescent round his neck. Finally he put on his headdress of parrot skins and bird-of-paradise feathers. Then he said to his wife and child, "Come close and wait."

The crocodile began to pull him down into the water. He took off his legbands and threw them ashore. The crocodile pulled him farther and he took off his shell girdle. The crocodile pulled him farther till the water came level with his armpits, and then he took off his armbands and threw them ashore. The crocodile pulled him farther, and finally he took off his mother-of-pearl crescent and his feather head-dress and threw them ashore. He said, "It is done", and then he said, "Go! my wife, my child, go! What is become of me?"

Then there was the sound of splashing, the crocodile waved its tail and bits of leaf and grass were stirred up from the bottom of the water.

CHAPTER XII

The Preferred Types

SHORTLY after my return from New Guinea I read for the first time Kretschmer's *Physique and Character* (English translation, 1925), and it was at once obvious that the contrast which I had observed between the sexes in Iatmul culture was in some ways comparable with Kretschmer's contrast between cyclothyme and one group of schizothyme personalities.[1] It appeared that schizothyme behaviour was "fashionable", standardised in Iatmul culture as appropriate to men; while cyclothyme behaviour was standardised for women. The equivalence between the schizothyme behaviour of the Iatmul men and that of certain natives of South Germany is only partial; and the exact analysis of the relationship between Iatmul schizothymia and German schizothymia requires a great deal more research and a knowledge of the possible standardisations of schizothyme ethos in other cultures.

In the case of the equivalence between the Iatmul women's

---

[1] I should perhaps state my opinion of Kretschmer's dichotomy. I believe that, though the external world is perhaps not built upon a dualistic basis, dualisms and dichotomies provide a convenient technique for describing it, and this technique is so standardised in our culture that there is little hope of avoiding it. I agree, of course, with Kretschmer that we must not think of these dichotomies as discontinuous, but rather that we should expect to classify individuals on a scale varying between the extremes. With further investigation, we shall probably be able to devise a number of other dichotomies, so that our classification of individuals will no longer be a matter of setting them in a row between two extremes. With every new dichotomy a new dimension would have to be added to our map of possible variations.

It is possible, too, that some of our future subdivisions will occur only among schizothymes or only among cyclothymes, and indeed it already seems to me that a number of different syndromes are confused under the general term "schizothyme", while the concept of cyclothymia with its good correspondence with Jung's "extravert" and Jaensch's "integrate" is more likely to be a unity.

Such classification is an important preliminary to an understanding of the working of the various systems of personality, and is especially useful in presenting problems of patterns of behaviour between personalities.

ethos and Kretschmer's cyclothymia, I must insist that the resemblance between these two syndromes is solely in terms of easy emotional acceptance and jolliness; and that I know of no hint of *periodic* variation between exuberant joy and depression in the women's ethos, such as is characteristic of cycloid personalities. It is doubtful whether such an individual tendency could be culturally standardised in such a way that all the individuals were "in step" with one another, i.e. all manic at one time and all depressed at another. More probably, if periodic changes were standardised in the women, each would have her separate periodicity, and it would be hard to demonstrate whether this periodicity was a product of cultural standardisation or an expression of individual physiological deviance.

My material, which was not collected with these problems in view, is not good enough to be the basis of an exact analysis of these various syndromes, but since the contrasting sex ethoses of Iatmul culture are at least reminiscent of the types described by Kretschmer, it is worth while to consider how far the Iatmul are typologically conscious. To what extent have they developed ideas about the association of physique with character, and what types do they recognise?

The natives regard two types of man with approval. The first is the man of violence and the second the man of discretion. Of these the violent type is the most admired, and such a man is described with enthusiasm as "having no ears". He pays no attention to what is said to restrain him but suddenly and recklessly follows his assertive impulses. Such a man is represented in Plate XXII, and indeed this man was a little too sudden and unstable even for Iatmul taste. They regarded him as somewhat "cranky" and warned me against him when I took him as an informant. In this capacity he proved more curious than useful—very enthusiastic, but too hasty and astonishingly inaccurate. He seemed indeed to lack all power of critical thought and to have no sense of logical consistency. When his contradictory statements were presented to him he had no realisation of their incompatibility.[1]

---

[1] The occurrence of such a cognitive trait in a man who was the extreme of the preferred affective type is especially interesting since the

Such men though admired would, I was told, not be trusted with esoteric information, because the natives fear that in the erudite debating about the system of names and totems, such an uncontrolled person may blurt out some important piece of secret lore or provoke a brawl by too rashly exposing his opponents' secrets. Thus with his little knowledge of esoterica, the violent man will behave in debate in the sort of way which I have described above, filling out his speeches with histrionics and obscene reference.

The more discreet type is, I think, generally heavier—more pyknic—in physique and quieter and rather more at ease in his public appearances. He it is who is the repository of mythological knowledge, and it is he who contributes erudition to the totemic debating and keeps the discussion on more or less systematic lines. His balance and caution[1] enable him to judge whether to expose his opponents' secrets or merely to indicate by some trifling hint that he knows the secrets, such a hint being tantamount to a threat of exposure. He knows how to sit quietly in the debate carefully watching his opponents to judge whether they really know any of the important secrets of his clan or whether their trifling hints are only a bluff to frighten him into ceding some point.

In mythology these two types are contrasted. There is a series of tales of two brothers of whom the elder, Kamwaimbuangga, was of the discreet type while the younger, Wolindambwi, was a man of violence. Of these it is the latter who is the great hero, but who in fits of temper set fire to the original mythological ceremonial house and killed his sister's son. This reckless hero is said to have been a man of great beauty and especially to have had a long nose which was much admired. It was also stated to me in Mindimbit that his patrilineal descendants, the members of the half-clan which claims him as ancestor, have inherited from him noses more beautiful than those of the descendants of his discreet

Iatmul culture itself contains so many "contradictory" formulations. Questions of the relationship between affective and cognitive standardisations will be considered in a later chapter.

[1] A cyclothyme observer would probably prefer to describe such a man as "well balanced", while to a schizothyme he appears "cautious" if he be a friend, or "sly" if he be an enemy.

elder brother. Among these beautiful descendants my informant counted Mwaim-nanggur, the mythological hero whom all the women loved and who was finally murdered by their jealous husbands. (Cf. *Oceania*, 1932, Plate VI, which shows a portrait skull of Mwaim with enormous nose.)

I was for a long time puzzled by the constant references to long noses and by the conventional exaggeration of the nose in artistic representations of the human face (Plate XXVIII A and B). But the matter was partly cleared up by a discussion of the beautiful woman's head represented in Plate XXV. This is the head of a woman who died young, about three generations ago. The face is modelled in clay upon her skull and the portrait is preserved as a ritual object for use as the head of a *mbwatnggowi* figure in certain ceremonies (cf. Plate XXVII). Among the heads used for these dolls, some were trophies of war, while others were the heads of villagers who had died in peace. I asked on what principle *her* head had been chosen, and the natives said that any head of special beauty might be picked for this purpose and pointed to the length of the leptorrhine nose as the conspicuously attractive feature.

Thus the exaggeration of the nose in Iatmul art is in part at least a conventional reference to the standard of beauty. But there is probably another factor at work.[1] The enormous noses are certainly to some extent phallic symbols. They may be lengthened downwards to join the penis or navel, or they may end free with a representation of the head of a snake or bird at the tip. I have indicated elsewhere that in this culture phallic symbols are to be regarded, not simply as symbols of the genital organ, nor as symbols of fertility, but rather as symbols of the whole proud ethos of the males. Thus the flutes, the secrets of initiation, are phallic in mythological origin and stand in the culture as an outstanding symbol of the differentiation of the male sex by drastic initiation.[2]

---

[1] I collected one instance only in which a small nose was regarded as desirable. This was in the bullying of the novices by a masked figure. Each novice was made to rub noses with the mask and to say: "Your nose is a small nose, mine is big (*nemwan*)" and to express a liking for the contact. Otherwise, I invariably heard praise for long (*tshivla*) noses.

[2] "Music in New Guinea", *The Eagle, loc. cit.*

In ideas about noses, the symbolism seems to have gone full circle. The schizothyme ethos of the men is linked with emphasis upon a leptorrhine standard of beauty, and the use of phallic symbols for the schizothyme ethos has involved the use of the leptorrhine nose as a phallic symbol.

Unfortunately I was not familiar with Kretschmer's typology when I was in the field and therefore did not enquire into any ideas which may be present in this culture as to the correlation of violent character with leptorrhine nose. The attribution of such a nose to Woli-ndambwi and its denial to the discreet elder brother, however, seem to indicate that the Iatmul have some vague idea of this sort.

It is interesting that the same type of nose is admired in women, in whom a violent proud temperament would seem at variance with their jolly co-operative ethos. But in spite of the contrast in sex ethos the men have a leptosome standard of beauty for both sexes. They prefer the physical type which Kretschmer maintains is associated with the schizothyme temperament, and articulately regret that Iatmul women tend to have "bad small noses".

I have already mentioned that the men feel a distaste for the women's ethos, and it seems that the same attitude is reflected in their standards of beauty. But while it is certain that the men admire the woman who recklessly takes the initiative in sexual affairs, even risking her life in the venture, it is by no means certain that the same type of woman is popular with her own sex. I did not collect any statements of the women's views on these matters. It is possible that the leptorrhine standard of beauty in women is linked with the occasional emphasis upon pride which is described in the chapter on the women's ethos.

In a very few men the admired qualities of both the violent and discreet types were combined, or so it seemed to me. Such a man was Mali-kindjin, who is now dead, but who was the greatest man in Kankanamun. Toravi of Angerman has something of the same greatness.

Mali-kindjin was an old man, and when I knew him he was an invalid. They said that his sorcery was beginning to recoil upon his own head. He was an astonishingly vivid and dramatic orator; when one entered his house he would come forward and make a speech of welcome, not unctuous but beautifully crisp and hard. In the initiatory system, he was a member of grade 2 (cf. Diagram, p. 245) and therefore nominally had no active part in the business of initiation, but he was always on the spot criticising the proceedings and correcting the gong rhythms. On one occasion the father of a novice was angry because something had been jabbed into his son's eye. Mali-kindjin took the side of the initiators and in the debate which followed he suddenly seized a log from the fire on the ground and with it belaboured the angry father—who took his punishment in silence.

Shortly before he died, Mali-kindjin was trying to change his own position in the initiatory system, to get himself demoted so that he might take part in the ceremonial. His father had caused him to be promoted when he was a boy from grade 6 to grade 4 (cf. Diagram, p. 245). He now claimed that this promotion had been irregular, and that he should now be a member, not of grade 2, but of grade 4, where he would play a more active part.

He was well hated and feared for his sorcery, and when it was clear that he was really ill, a debate was organised against him. His esoteric claim to the Sepik River as a clan ancestor was attacked. He told the members of his own clan to keep quiet during the debate— he would do all the speaking himself. After four hours of acrimonious debate and insult, continually dancing his scorn of those who would impugn *his* Kindjin-kamboi (the snake which is the Sepik River, according to his clan's mythology), he deliberately exposed an esoteric secret of the opposition. He picked up a yam tuber and began to dance with it in his mouth—a reference to the opposition's secret myth of the origin of flutes from a yam. A brawl started at once and Mali-kindjin got some nasty knocks and had a stool thrown at his head. The debate was resumed and went on till evening, Mali-kindjin still keeping up his flow of oratory. Then, at dusk, when it was all over, he went and sat alone, an exhausted and sick man, at the end of the ceremonial house, and began to chant the name songs of his clan—his eyes shining.

The feelings of the other natives about Mali-kindjin were very definitely ambivalent. For example, Tshava, his own sister's child, had in his boyhood a great deal of contact with Mali-kindjin, and even took part in some of his sorcery. Mali-kindjin wanted to teach Tshava his spells, but the latter was frightened. He says that he was frightened lest Mali-kindjin should take offence at the small payment for the secrets, but I suspect, too, that he was frightened of the dangerous position in which a sorcerer finds himself, hated by the other members of his village. In any case, he says that it was partly on the advice of his father that he did not associate more with Mali-kindjin, and now he regrets this and says: "Now me stop along bloody fool, that's all." The "bloody fool" is his father.

When Tshava discussed Mali-kindjin's doings with me, he quite clearly admired the latter for his breaches of convention. It is usual to present shell valuables to the medicine man when his services are engaged: but these valuables, though nominally they become the property of the medicine man, are usually returned to the donor. The patient when he is cured brings to the medicine man a bunch of areca nuts and places these upon the latter's shoulder. The medicine man then returns the valuables which had been given to him. Tshava told me with glee that Mali-kindjin always kept both the valuables and the areca nuts.

Mali-kindjin's son, Avuran-mali, had also a very great admiration for the old sorcerer and pride in his doings. After the debate mentioned above in which Mali-kindjin was badly knocked about, Avuran boasted to me of how his father had won the debate—how he had exposed his opponents' secrets and thereby exasperated them into a brawl.

Other people, however, not so closely related to the old sorcerer, regarded him with frank detestation. It was pointed out to me as evidence of his sorcery that "he had no relatives", the inference being that he had killed his own relatives to prevent the sorcery from recoiling upon his own head. But though they feared him, even hated him, they were proud, too, that he was a member of their village, and recognised that Mali-kindjin in the old days had helped them both magically and physically in warfare.

Another type of man who is recognisable in every village is the sorcerer who lacks violence—skinny, pinched individuals who have compensated for their lack of spirit by cultivating a reputation for skill in magic. Such a man was Namwio of Mindimbit, a henpecked husband for whose wife everyone was sorry; and Tshaun-awan of Palimbai, a man who had never married, was another of this type. Such men are useful; they perform cures and, as public servants, carry out the magic to cause the water level to rise and fall when required. They also may be paid to cause death and sickness or to raise the water level in the interests of some individual at a time when the majority would benefit by a fall. I think that in the past little or nothing was done to punish these activities, but nowadays there is a tendency to try to persuade the District Officer to put these men in prison as scapegoats for the vagaries of the floods. As informants I have come in contact with three of these men, and in each case I have been surprised to find that their esoteric knowledge is actually very limited. They cannot really recite their spells, but only jumble them. The native comment upon these men is—"they have no meat on their bones."

Another insult which is levelled at personalities whom the natives do not admire is the phrase: "*kau tapman nyan*" (literally "fight—none—child", i.e. a man with no fight in him).

There is another line of enquiry into native personalities which I did not pursue but which, I believe, would have given positive results. Though my material is inconclusive, it is perhaps worth giving since it presents a new aspect of the problems of culture contact. At the present time, the villages of the Iatmul contain considerable numbers of young men who have recently returned to their homes after spending from three to five years as indentured labourers on European plantations and gold mines. They grew up as boys in the Iatmul ethos probably admiring it and probably believing that the ethos was "natural" for men. Then they went away and lived for some years in the more disciplined and co-operative ethos of a labour line. Now they are returning to the ethos in which they grew up.

I believe that among these young men, when I recall their personalities, I can roughly discriminate those to whom the schizothyme behaviour was natural or innate, and that such young men when they come home drop easily back into the native community. But the others appear to have been born with or to have acquired but little bent for schizothyme behaviour. As boys, they probably adopted the prevailing ethos because it was fashionable and was the only pattern offered to them. But now, when they have seen and lived in a different ethos, these men look askance at their native culture. They are impatient of the buffooning of the older men and they treat the most important rituals with contempt. They are openly careless about the secrets of initiation.

An anecdote will illustrate the sort of heresy which is now growing in the community as a result of the culture contact:

The men ceremonially concluded an elaborate *wagan* ceremony which had lasted for many weeks. During all that time[1] the *wagan*, the sacred initiatory slit-gongs, had been beaten and there had been a strict taboo on the use of the secular gongs. At the close of the ceremony, the men, in ritual procession, put away the sacred heads of the two *wagan* which had been exhibited in front of the gongs during the ceremony and then they beat a special rhythm on the secular gongs, thus removing the taboo.

In the very early morning of the following day, I was woken by some men who came to my bedside and spoke somewhat as follows: "We want to kill Tshimbat's pig. Formerly we would have killed *him*, but now we are afraid of prison. If we kill his pig will the Government put us in prison? Yesterday he and the other boys in the toy ceremonial house carried out a mock ceremony in imitation of ours. They beat the rhythm before we did. They made our ceremony futile."

Tshimbat (Plate XXIV B) was a very bumptious, noisy, somewhat pyknic, returned labourer, whose house the men had had to raid some months previously at dead of night, prodding the house with long bamboos to overturn and smash the pots, because Tshimbat had been fooling with a

[1] Cf. footnote, p. 137.

hand-drum when he ought not. I knew him well as a disturber of the peace and knew that the Government would support what was fine in the native culture. I said: "My shoot boy shall take my gun and shoot the pig." This was done and we all ate it.

That day Tshimbat's mother came out onto the dancing ground and stood just in front of the ceremonial house, dirging for the pig which she had raised. A fine schizothyme old man, Djuai (Tshava's father), seized a stick and beat her thoroughly, driving her off.

But the dice are heavily loaded in favour of the discontented and maladjusted young men, and a further interference from me was necessary before my conservative friends were finally successful. A few days later, entering the ceremonial house one morning, I was surprised to see a row of currency shells set out as if a subscription was being raised for some feast. I asked what feast was in the air, and was told: "No. Tshimbat has been talking. He is going to make a court case of it. We shall all go to prison. He knows pidgin English and we don't. We are going to pay for the pig." Then I made a speech at the debating stool and said that the only man who was legally liable was Djuai and that the shells should be put in a basket and given to him if he suffered; that the pig had been killed by my shoot boy with my gun and acting on my orders and that under no circumstances would I give anything to Tshimbat. So the shells were put in a string bag, and we all waited for the District Officer.

When the case was heard, Tshimbat spoke querulously of the old men, stated a preference for the customs of the white men, and received a very sharp snub for his pains. Djuai said little, but stood puzzled, nervous and rather stiff. He was fined ten shillings for assault. No damages were awarded to Tshimbat and his mother.

Then we withdrew to the ceremonial house to consider what should be done with the bag of currency shells. The natives dislike changing shells into shillings and it was therefore decided that the shells should be returned to the donors and, without interference from me, a fresh collection of shillings was started. I contributed one shilling, and four more were added. Then Djuai came forward to the debating stool. He said that he had wanted for other reasons to hit that woman. She had been unkind to his child, and he would be well content with five shillings.

In this instance I was able to defend the culture, but without my interference it is almost certain that the initiatory system in Kankanamun would have received a severe blow. I suggest that a very great deal of the action of culture contact in destroying institutions may be due to mechanisms of this kind—the contact upsetting the delicate adjustment between the temperaments of deviant individuals and the ethos of the culture.

CHAPTER XIII

Ethological Contrast, Competition *and* Schismogenesis

THE foregoing description of the contrast between the men's and women's ethos among the Iatmul people raises at once the problem of how such ethological contrast is produced and maintained. The first theory which occurs to us is that individuals of the two sexes, owing to genetic and deep physiological differences, may tend to develop different patterns of personality, and that the differences in ethos in the two sexes are simply an expression of these innate differences. But though we invoke heredity we cannot exclude the influence of culture and environment.

The matter is confused and difficult. The two extreme theories, (a) that the ethological contrast is entirely determined by culture and (b) that the contrast is entirely determined by sex physiology and heredity, are both of them untenable. The first must be abandoned because we know of considerable innate physical and physiological differences between the sexes, differences in shape, in bulk, in deposition of fat, in speed of development and so on. These differences must necessarily be expressed in all the behaviour of the individuals—in gesture, posture, choice of activity, etc.—and therefore must contribute to the ethological contrast.

On the other hand, the theory which would ascribe the contrast entirely to heredity must be abandoned for various reasons. Even if we were dealing with physical differences between the sexes in any one community, the question would arise as to how much these physical differences are culturally exaggerated or modified by costume or deformation; and the likelihood of cultural modification of psychological differences is even greater.

Some theory of a fundamental biological difference in temperament between the sexes may probably be found in every community in which the ethoses of the two sexes are

differentiated, and in our own culture with its extreme confusion and diversity of ethos, it is usual to ascribe almost any
characteristic of a personality to biological sex. Under the
term "womanliness" is subsumed a syndrome of characteristics—warm motherliness, easy emotional expression,
pyknic beauty and so on—comparable with the ethos of
women among the Iatmul. But the term "femininity" is used
for a syndrome much more closely allied to Iatmul-masculine-
pride, sudden capriciousness, waywardness and leptosome
beauty.

More significant, however, than the confused phrasings of
our own culture are the findings of Dr Margaret Mead.[1] She
has shown that among the Arapesh who live in the mountains
north of the Sepik River, between it and the coast, a uniform
ethos is standardised for both sexes; and that among the
Mundugumor who live on a tributary of the Lower Sepik,
there is again no contrast in sex ethos. But while the ethos
of both sexes among the Arapesh, though more gentle, is
vaguely reminiscent of that which we have found among the
Iatmul women, the ethos of both sexes among the Mundugumor is a harsher and less exhibitionistic version of that of the
Iatmul men. Lastly, Dr Mead worked among the Tchambuli
and found there a contrast between the ethos of the two sexes.
This contrast was, however, not identical with that which
I have described among the Iatmul. The ethos of the Tchambuli men was less harsh and more exhibitionistic than that
of the Iatmul, while the Tchambuli women were somewhat
harder and more business-like than those of the Iatmul. In
the light of these findings, if we are to maintain a theory of
innate difference in temperament in the two sexes, we should
have to suppose that the genetic constitution of the individuals is, statistically, markedly different in each of the four
tribes which I have mentioned.

Inasmuch as there are physical differences between these
peoples and a likelihood that these physical differences are
genetically determined, it is conceivable that there may be
differences in innate temperament. But at least we can say
that the differences are not due to peculiarities of sexual

[1] *Sex and Temperament*, 1935.

physiology, since either ethos may be standardised for both sexes. The differences, whether sexual or not, would have to be described in Mendelian terms, and I know of no pattern of Mendelian inheritance which would enable us to build up populations with prominent statistical differences or similarities between the sexes such as occur in the four tribes which have been described.

We are thus forced to take up some position intermediate between the two extreme theories. We cannot entirely exclude either heredity or social environment, and it would be premature to indicate the exact form of intermediate theory which we should adopt. In this chapter, since the material I collected was not of a sort to be susceptible of genetic analysis, and since I took no measurements of physique, I must confine myself to the investigation of the social and cultural factors involved in the shaping of personality, while leaving open the possibility that such temperamental characteristics as are referred to in the terms, cyclothymia and schizothymia, may be determined by heredity and are probably independent of sex.

It is possible, for example, that the populations in the four tribes mentioned above are statistically alike in temperamental characteristics, and that in each population and in each sex there occur individuals who are born with a natural bent for the various ethoses. If this could be demonstrated, we should then have to suppose that some of the individuals in any one culture are naturally more fitted for life in that culture than others, and that in each culture there are genetic deviants who do their best to adapt themselves to an ethos whose bias is but little developed in their temperaments. We should, however, still have to acknowledge that hereditary tendencies had played a very important part in the shaping of the culture since the innate characteristics of the "preferred types" would seem to have guided the culture in its evolution.

With this preamble, we may turn to consider cultural factors which probably promote the ethos of each sex in Iatmul culture. When I was among the Iatmul, the very concept of ethos was to me very dimly defined, and I have never had any training or practical experience in the study of the moulding of human beings and especially children. I have therefore not the material to describe these processes among the Iatmul in any detail, and my remarks must be regarded as no more than a tentative suggestion of how the processes may fit together.

My impression is that there is no marked difference between the treatment of male babies and that of females, nor did I find any strong feeling that babies of one sex were more desirable than those of the other. In general, infants of both sexes seem to be happy and well-treated; they are, I think, rarely left alone for long periods. One detail is worth mentioning: in this culture children of both sexes are conspicuously ornamented with shells even in everyday life (cf. Plate XVII A).[1] I used to carry on my wrist a circular pig's tusk (bought in Sydney) and found that the loan of this shiny white object, obviously an ornament, was invariably effective in stopping the crying of small children of both sexes, at least for a few minutes.

It is my impression that we should look for the origins of contrasting sex ethos among the Iatmul, not in the experiences of very early childhood, but in the later training of boys and girls. We should see the two ethoses as acquired by learning and imitation rather than as springing from peculiarities implanted in the deep unconscious in the first two years of life.

In the case of the boys: the preoccupation of their seniors with head-hunting and with the production of spectacular displays; their life in the toy ceremonial house, where they ape their seniors; the processes which we have analysed in our study of initiation (p. 130 *et seq.*); the little boy's first experience of homicide when he is still a child, spearing some wretched bound captive, while his *wau* helps him to lift the spear; the elaborate *naven*, of which the little boy is immediately made the hero—all these factors, no doubt, contribute to shape the boys into Iatmul men.

---

[1] It is possible that this ornamentation of children is in some sense analogous with the ornamentation of women, which is a conspicuous characteristic of the Eastern Iatmul (cf. footnote, p. 142).

Similarly, the women's ethos is no doubt formed in part by their preoccupation with the routines of food-getting and child-rearing, and by the association of girls with older women who have already adopted the ethos.

Factors of the sort which I have considered above may be presumed to play their part in maintaining the *status quo*, but in addition to these I believe that in the mechanisms which underlie ethological contrast, we are concerned with other factors which, if they were unrestrained, would lead to changes in the cultural norms. I am inclined to see the *status quo* as a dynamic equilibrium, in which changes are continually taking place. On the one hand, processes of differentiation tending towards increase of the ethological contrast, and on the other, processes which continually counteract this tendency towards differentiation.

The processes of differentiation I have referred to as *schismogenesis*.[1] They are, I believe, of very wide sociological and psychological significance, and therefore in my description of these phenomena I shall use as illustrative material not only the very meagre facts which I collected in New Guinea, upon which the concept of schismogenesis was originally built, but also no less sketchy observations of the occurrence of schismogenesis in European communities.

I would define schismogenesis as *a process of differentiation in the norms of individual behaviour resulting from cumulative interaction between individuals*.

For the moment we need not concern ourselves with defining the exact position of this concept in regard to the various disciplines which I have tried to separate. I think that we should be prepared rather to study schismogenesis from all the points of view—structural, ethological, and sociological—which I have advocated; and in addition to these it is reasonably certain that schismogenesis plays an important part in the moulding of individuals. I am inclined to regard the study of *the reactions of individuals to the reactions of other individuals* as a useful definition of the whole discipline which

---

[1] *Man*, 1935, p. 199, "Culture Contact and Schismogenesis". This article is an outline of the sociological problems presented by the concept of schismogenesis. It contains almost no reference to the phenomena which I observed in New Guinea.

is vaguely referred to as Social Psychology. This definition might steer the subject away from mysticism.

We should do well, I think, to speak no more of "the social behaviour of individuals" or of "the reactions of the individual to society". These phrasings lead all too easily to such concepts as those of Group Mind and Collective Unconscious. These concepts are almost meaningless to me, and I believe that even if we avoid them we are liable to err by confusing our study of the psychological processes of the individual with our study of society as a whole; a confusion of spheres of relevance.

When our discipline is defined in terms of the reactions of an individual to the reactions of other individuals, it is at once apparent that we must regard the relationship between two individuals as liable to alter from time to time, even without disturbance from outside. We have to consider, not only A's reactions to B's behaviour, but we must go on to consider how these affect B's later behaviour and the effect of this on A.

It is at once apparent that many systems of relationship, either between individuals or groups of individuals, contain a tendency towards progressive change. If, for example, one of the patterns of cultural behaviour, considered appropriate in individual A, is culturally labelled as an assertive pattern, while B is expected to reply to this with what is culturally regarded as submission, it is likely that this submission will encourage a further assertion, and that this assertion will demand still further submission. We have thus a potentially progressive state of affairs, and unless other factors are present to restrain the excesses of assertive and submissive behaviour, A must necessarily become more and more assertive, while B will become more and more submissive; and this progressive change will occur whether A and B are separate individuals or members of complementary groups.

Progressive changes of this sort we may describe as *complementary* schismogenesis. But there is another pattern of relationships between individuals or groups of individuals which equally contains the germs of progressive change. If, for example, we find boasting as the cultural pattern of

behaviour in one group, and that the other group replies to
this with boasting, a competitive situation may develop in
which boasting leads to more boasting, and so on. This type
of progressive change we may call *symmetrical* schismogenesis.[1]

In the light of this theoretical consideration of the possibilities, it is clear that we must consider the various contexts
of Iatmul culture to determine whether either complementary
or symmetrical schismogenesis contributes anything to the
shaping of cultural norms. Is it possible that the contrast
in sex ethos is of such a kind that it is liable to complementary
schismogenesis?

We have seen that the women are an audience for the
spectacular performances of the men, and there can be no
reasonable doubt that the presence of an audience is a very
important factor in shaping the men's behaviour. In fact,
it is probable that the men are more exhibitionistic because
the women admire their performances. Conversely, there can
be no doubt that the spectacular behaviour is a stimulus
which summons the audience together, promoting in the
women the appropriate complementary behaviour. We may
wonder, too, whether the whole system of behaviour which
surrounds the flutes, *wagan*, and other secrets of initiation
would be maintained if it were not for the fact that the women
hear and admire the music of the flutes and the rhythms of
the *wagan*.

But the contrast between exhibitionism and admiration is
only a part of the general sex contrast which includes a
whole nexus of inter-related characteristics, and it would be
very important to know whether this wider contrast tends
towards schismogenesis. The only detail which I can produce which seems to show that such a schismogenesis occurs
in Iatmul culture is the reaction of the men to the dirging

[1] The difference between complementary and symmetrical schismogenesis is closely analogous to that between schism and heresy, where
heresy is the term used for the splitting of a religious sect in which the
divergent group have doctrines antagonistic to those of the parent group,
while schism is the term used for the splitting of a sect in which two
resulting groups have the same doctrine, but separate and competing
politics. In spite of this, I have used the term schismogenesis for both
types of phenomena.

of the widow, when they indulge in harsh caricature. Unfortunately I do not know the widow's reaction to being caricatured.

Complementary schismogenesis is evident again in the contexts of initiation, the process of inculcating Iatmul-masculine ethos into the novices. This process I have already analysed in such detail as the material permits (p. 133). I have indicated how the initiators with Iatmul-masculine ethos react to the presence of novices with (supposedly) Iatmul-womanly ethos: the presence of the novices stimulates the initiators to harsh and fantastic behaviour. Later, the novices themselves become initiators, and are schismogenically driven by the presence of later novices into Iatmul-masculine behaviour. The admiring ethos of the women is also active in bringing about the assimilation of the novices to the group of men; the novices are ornamented and exhibited to the women at the end of initiation.

Lastly, a curious and possibly factitious case of complementary schismogenesis may be noted in the *laua*'s boasting in the presence of his *wau*, who shames himself in reply.

Symmetrical schismogenesis is not evident between the sexes, but occurs in dramatic form in initiation. Here we have rival moieties competing against each other in their bullying of the novices, and prompting each other to further brutalities (pp. 134, 135).

The phenomena of schismogenesis are by no means confined to Iatmul culture, and in order to emphasise the widespread importance of the process which I first noted among the Iatmul, I shall indicate in what other fields I expect to recognise schismogenesis:

1. In all intimate relations between pairs of individuals. A great many of the maladjustments of marriage are nowadays described in terms of the identification of spouse with parent. Such a phrasing may be historically accurate: it may be true that the husband in a marriage tends to carry over into his relationship with his wife attitudes which have been previously formed in his relationship with his mother. But this fact alone is by no means sufficient to account for the break-

down of the marriage, and it is difficult in terms of such a theory to explain why such marriages, in their earlier stages, are often very satisfactory and only later become a cause of misery to both partners.

But if we add to this diachronic phrasing of the relationship the possibility that the patterns of behaviour between the partners are liable to progressive change of a schismogenic nature, it is evident that we have a theory which would explain both why the relationship is satisfactory in its early stages and why its breakdown appears inevitable to the people concerned. The relationship between son and mother is, in our culture, a complementary relationship,[1] which in its early stages is patterned on fostering on the mother's side and feebleness on the child's. Later the relationship may develop in very various ways: e.g., (a) the pattern of fostering and feebleness may persist; or (b) the relationship may evolve towards the pattern which we have noted among the Iatmul where the mother takes a vicarious pride in her offspring; or (c) it may evolve towards an assertive-submissive contrast in which it may be either person who takes the assertive role. But whatever the pattern, the mother-son relationship is almost always complementary. If these patterns are carried over into the son's marriage it is likely that they may there become the starting point of a schismogenesis which will wreck the marriage.

It is perhaps worth while to suggest that in many cases an explanation and demonstration, to the partners in such marriages, of the schismogenesis in which they are involved might have the same therapeutic effect as the understanding, on the part of the husband, that he is identifying his wife with his mother.

2. In the progressive maladjustment of neurotic and prepsychotic individuals. I myself have no experience of psychiatry, but I suspect that in addition to studying the

---

[1] Exceptionally, in our own culture, we may come across cases in which the relationship between son and mother is almost symmetrical or reciprocal. It would be interesting to know whether such sons are ever, in their marriages, involved in. complementary schismogenesis of the fostering-feebleness or exhibitionism-admiration types.

individual pathology in every case, the psychiatrist would do well to pay more attention to the relations which the deviant individual has with those around him. I have discussed this matter with Dr J. T. MacCurdy, who agrees with me that in many cases the growth of the symptoms of the paranoid individual are attributable to schismogenic relationships with those nearest to him. I understand that it is usual to find that those paranoids who build their delusions around a belief in the unfaithfulness of their wives, almost invariably have wives whose utter faithfulness is obvious to every outsider. Here we may suspect that the schismogenesis takes the form of continual expression of anxiety and suspicion on the husband's side, and continual response to this on the side of the wife, so that she, either continually humouring him or contradicting him, is promoting his maladjustment, and he, in turn, becoming more maladjust, demands more and more exaggerated responses from her.

In the case of schizoid maladjustment, the matter is not so clear. I have suggested above that in Iatmul culture the circular ethos of the women and the schizothyme ethos of the men are mutually complementary and liable to schismogenesis. If this be true, and further observations are required to verify it, we must be prepared to accept the fact that the schizophrene is not merely working out his own internal pathology, which indeed may or may not be getting worse, but is also responding to the more cyclothyme people around him by himself becoming more and more schizoid.

Such a view of the progressive degeneration of the schizoid would explain very simply the extraordinary inevitability of this degeneration, and would explain, too, the fact that the patient himself is often preoccupied with ideas of predestination in such a degree that these ideas themselves contribute in no small measure to his destruction. It is possible that if in the early stages a schizoid individual could be brought to realise that the progressive process which he sees as inevitable is actually very simple and easily controlled, its progress might be arrested.

But a great deal of work remains to be done in labelling the various syndromes which are grouped under the heading

of "schizothymia". From the Iatmul material, it is clear
only that the contrast between the men and the women is
somehow comparable to that between one type of schizothymia and the circular temperament. It is probable, however, that there are a good many patterns of complementary
schismogenesis in each of which one partner is ultimately
driven into schizoid distortions of his personality, and that a
classification of these various paired patterns will give us a
clue to the classification of the schizothyme syndromes. Moreover, an understanding of these syndromes in terms of the
schismogeneses in which they arise may give us a clue to
their treatment in the early stages.

In the later stages of schizophrenia it seems probable that
the personality of the patient is permanently maimed, and
an understanding of the schismogenesis which contributed
to his breakdown would probably be of no therapeutic use.

It is probable that schismogenesis is an important factor
in neurosis as well as in psychosis, and that a new discipline
of psycho-analysis could be built up on these lines supple-
menting the systems which are now being used. In Freudian
analysis and in the other systems which have grown out of it,
there is an emphasis upon the diachronic view of the individual, and to a very great extent cure depends upon inducing
the patient to see his life in these terms. He is made to realise
that his present misery is an outcome of events which took
place long ago, and, accepting this, he may discard his misery
as irrelevantly caused. But it should also be possible to make
the patient see his reactions to those around him in synchronic
terms, so that he would realise and be able to control the
schismogenesis between himself and his friends.

It is likely that while for some patients the administering
of a diachronic view is curative in its effects, for others this
treatment may only accentuate their maladjustment. For these
latter, it is possible that the administering of a synchronic
view would be curative and give them a complete and realistic
understanding of themselves. The dangers of psycho-analysis
when administered to schizoid patients may indeed arise
simply out of the preoccupation of such patients with destiny
and the inevitability of historic accident. A sense of con-

temporary process is perhaps a necessary corrective to an over-developed sense of personal history, and *vice versa*.

I believe also that it would be worth while to look for schismogenic phenomena in cases of split personality. Here we have two forms of schismogenesis to consider: first a probable schismogenesis between the patient and his friends, and secondly a possible schismogenesis *within* the personality of the patient. It is possible, in fact, that one half of the split personality promotes the other and *vice versa*, producing an ever-widening breech and incidentally causing each half of the personality to be less and less capable of adaptive behaviour in the patient's social setting.

To what extent this schismogenesis within the personality is present in all individuals who are involved in an external schismogenesis is a question which cannot now be answered, but it is probable that we may see the development of vicarious pride and its expression in gestures of self-abnegation as phenomena related to those of split personality. It is probable too that we should see exhibitionism and narcissism as very closely inter-related phenomena. The Iatmul man, who courts the admiration of the women by exhibitionistic behaviour, is almost certainly admiring his own performance with one half of his personality: an external schismogenesis may induce an internal narcissism.

Dr MacCurdy has suggested two other contexts in which schismogenesis probably occurs, namely in the behaviour of manic patients, and in the tantrums of children. The symptoms of mania are enormously increased when an audience is present; and the responses of parents are liable to accentuate the tantrums of their children.

It is not enough, however, to make a list of contexts in which schismogenesis occurs, and it would be a mistake to suppose that in the word schismogenesis we have the key to all the processes of character formation. If this concept is to be of real value, it must stimulate enquiry into the conditions upon which schismogenesis depends. Among these we may note that schismogenesis is impossible unless the social circumstances are such that the individuals concerned are held together by some form of common interest, mutual dependence, or by their social status. As to the nature of the factors which hold them together, and the relationship between these factors and the schismogenic process, we know nothing.

Another factor which is necessary for schismogenesis has already been mentioned but it assumes a special importance in these psychological contexts. I said (p. 176) that if the behaviour of A "is *culturally labelled as* an assertive pattern, while B is expected to reply to this with what is *culturally regarded as* submission", we may expect schismogenesis to occur. The ethological aspect of the behaviour is fundamental for schismogenesis, and we have to consider not so much the content of the behaviour as the emotional emphasis with which it is endowed in its cultural setting.

If A gives B some object, and B receives it, these acts may be seen, according to their setting, as: (*a*) a triumph for A, an attitude which may lead either to symmetrical schismogenesis of the "potlatch" type or to some form of complementary schismogenesis if the asymmetry is continued; (*b*) a triumph for B, an attitude which may lead to some schismogenesis on the general lines of commercial rivalry, or (*c*) a triumph for neither. Both giving and receiving may be seen merely as expressions of mutual friendliness, and their continuance may lead, not to schismogenesis, but rather to closer union between A and B.

When we are dealing with schismogenesis between two separate individuals or groups of individuals, it is clear that we must consider ethological emphasis as a necessary condition for schismogenesis. But the question becomes more difficult when we consider schismogenesis *within* a single personality. It is probable that some condition analogous to "ethological emphasis" must be postulated for such internal schismogeneses, but it is not clear how this is to be phrased without unduly personifying the separate elements in the multiple personality: a problem which must be left to the psychiatrists.

3. In culture contacts. I have already given an outline of how considerations derived from the concept of schismo-

genesis should be applied to the study of culture contact.[1] In my opinion, we should see the phenomena of contact as a series of steps starting from a point at which two groups of individuals, with entirely different cultures in each group, come into contact. The process may end in various ways, and the theoretically possible end-results of the process may be enumerated: (1) the complete fusion of the two groups, (2) the elimination of one or both groups, (3) the persistence of both groups in dynamic equilibrium as differentiated groups in a single major community.

At the beginning of the contact, however, at least where the contact is one between Europeans and "primitive" people, we cannot expect to find any simple relationship between the ethos of the one group and that of the other. But I suspect that in quite a short time the individuals of each group adopt special norms of behaviour in their contacts with individuals of the other group, and that these special norms of behaviour will be classifiable in terms of complementary or symmetrical patterns.

This phenomenon is clearly seen in New Guinea in the behaviour which Europeans adopt towards natives, and in that which the natives adopt towards Europeans. Several hundreds of different cultures, with the greatest diversity of ethos among them, have been concerned in this culture contact, and yet one fact strikes the anthropologist as soon as he reaches the country: the average European resident believes that the natives of New Guinea are remarkably similar in all parts of the country. The resident will grant that some are better workers, and that others have more courage, but broadly his opinion is that one "coon" is very much like another. The basis for this opinion is, I believe, to be found in the special behaviour which the natives adopt in their dealings with Europeans. Apart from slight differences, the majority of peoples of Mandated New Guinea have adopted the same tactics in dealing with Europeans, and they have no doubt copied each other in their methods.

Pidgin English, the language which has developed in this culture contact, and the system of gestures and intonation

[1] *Man*, 1935, p. 199.

which are almost a part of pidgin, seem to me to be an expression of a very definite ethos. The language has its own cadence and special flexibility, which we may suppose adapted to humour the sometimes assertive and always incalculable European. I am inclined even to think that the ethos expressed in pidgin English is comparable with that which is caricatured in American "negro-humour".[1]

In any case, it is certain that for a majority of the natives of New Guinea the ethos of pidgin contrasts sharply with the ethos of their native cultures. This was conspicuously true of the Iatmul, and the behaviour of returned labourers in the ceremonial house, etc., was a continual source of shock to me. I felt that it was vulgar while the behaviour of the other natives, though it might be harsh and noisy, had never this peculiar quality. I think that the natives who had never worked on plantations felt shocked in the same way.

The process, however, by which two groups whose respective cultures are mutually irrelevant evolve a complementary or symmetrical relationship in terms of behaviour which is normal to neither group, has never been investigated. It should, if susceptible of study, give us clues to the process of establishment of schismogenic pairs.

Once complementary patterns of behaviour have been established, it is my opinion that subsequent schismogenesis is responsible for many of the antipathies and misunderstandings which occur between groups in contact. Hand in hand with the ethological divergence, we find the development of structural premises which give permanence and fixity to the split. But it is not clear to what extent these formulations, of "colour-bar", "racial antipathy", and mutual avoidance, contribute to promoting the schismogenesis. It is possible that some of these formulations prevent it

---

[1] Dr MacCurdy has pointed out to me an interesting detail in the schismogenesis between negro and white in America—that the Southern negroes in a highly differentiated schismogenic relationship with the Southern whites are very unwilling to accept from Northern whites behaviour patterns they would expect and demand from the whites whom they know. Thus the patterns of behaviour developed in a schismogenesis cannot easily be applied in other contexts, and for an understanding of the equilibrium we shall need to know a great deal more than the mere details of schismogenic behaviour.

from going too far. These are problems which need investigation.

Ultimately the two groups reach one of the end results
which I have listed above. Of these, we may ignore the first
two and consider only the persistence of both groups in
dynamic equilibrium, a relationship between the groups in
which the tendencies towards schismogenesis are adequately
restrained or counteracted by other social processes. This end
result is closely analogous to the state of affairs which exists
in a primitive community between any two differentiated
groups, and it is likely that a study of such equilibria as those
between the initiatory moieties or between the sexes in Iatmul
society will contribute to our knowledge of these equilibria,
and will enable us to handle the problems of culture contact
in a more understanding manner.

4. In politics. In the present disturbed and unstable state
of politics in Europe, there are two schismogeneses which
stand out conspicuously: (a) the symmetrical schismogenesis
in international rivalries, and (b) the complementary schismogenesis in "class-war". Here again, as in the other contexts
in which we have discussed schismogenesis, the progressive
evolution of behaviour towards greater and greater differentiation and mutual opposition is evident, and our politicians
would appear to be as incapable of handling the process as
the schizoid is of adjusting himself to reality.

A third type of schismogenesis is of some local importance,
namely the process whereby dictators are pushed towards a
state which, in the eyes of the world, seems almost psychopathic. This is a complementary schismogenesis between the
dictator on the one hand, and his officials and people on the
other. It illustrates very clearly how the megalomaniac or paranoid forces others to respond to his condition, and so is auto-
matically pushed to more and more extreme maladjustment.

These vast and confused schismogeneses are played out amid
such complicated circumstances that it is scarcely possible
to study them. But it might conceivably be worth while
to observe to what extent in their policies politicians are
reacting to the reactions of their opponents, and to what

extent they are paying attention to the conditions which they
are supposedly trying to adjust. It may be that when the
processes of schismogenesis have been studied in other and
simpler fields, the conclusions from this study may prove
applicable in politics.

THE PROGRESS AND CONTROL
OF SCHISMOGENESIS

In the absence of any carefully collected material which
would show the progress of schismogenesis in any one case,
I can only offer my opinion as to what stages are likely to
occur. In the first instance, when the complementary or
symmetrical relationship is first established, it is probable
that the patterns of behaviour which the two individuals or
two groups adopt seem to both parties to be a satisfactory
answer to a difficult problem of relationship. Such at least
would seem to be the case, for example, in the adoption of
pidgin English ethos by the natives of New Guinea, and
probably the same is true of the early stages of schismogenic
marriage. In the case of psychological maladjustments, however, I must leave it to the psychiatrists to decide whether
the initial adoption of the complementary pattern is an adaptive process.

If, however, a complementary or symmetrical pattern,
when once adopted, becomes more and more emphasised by
schismogenesis, it is likely that the personalities of the individuals concerned will undergo some sort of distortion with
over-specialisation in some one direction, whether it be exhibitionism, fostering, assertion, submission or what not.
With this distortion some degree of discomfort will be
introduced into the relationship, and it may even happen,
though this requires verification, that the individuals in trying
to find again the answer which was formerly satisfactory,
actually specialise even further in their respective roles.

Sooner or later, the distortion of the personalities is likely
to be accompanied by three effects: (*a*) a hostility in which
each party resents the other as the cause of its own distortion,
(*b*) at least in complementary schismogenesis, an increasing

inability to understand the emotional reactions of the other party, and (c) mutual jealousy.

At a comparatively late stage in the schismogenesis, when the personalities of the members of the two groups have definitely begun to suffer from distortion, it is probably usual to find a development of mutual envy. The distortion is a progressive specialisation in certain directions and results in a corresponding under-development of other sides of the personality. Thus the members of each group see the stunted parts of their own affective life fully developed—indeed over-developed—in the members of the opposite group. It is in such situations that mutual envy arises. Not only do the serfs envy the aristocrats, but the latter develop a distaste for their own ethos and begin to crave the simple life.

It is not easy to judge how much of this mutual envy is developed between the sexes in Iatmul culture. Certainly the women take a very real pleasure in the adoption of male clothes and ways in the *naven*, and this factor—mild envy of the masculine ethos—may be regarded as an important motive in determining their transvesticism. But in the men, the corresponding envy cannot be detected. Outwardly, at least, they despise the womanly ethos, but it is not impossible that they have some unacknowledged envy of it. Their own ethos would not, in any case, permit them to acknowledge that there was anything to be said for the attitudes of the women and any envy which they may have they might well express in scorn of women—dressing in women's clothes for the purpose.[1]

[1] I am here indebted to Dr Karin Stephen, who on the basis of her psycho-analytic experience in Europe suggested this possibility to me—that unconscious jealousy of the women's ethos may be one of the motives of the men's transvesticism.

I am very doubtful whether such a phrasing of the matter is permissible. I have endeavoured to use references to emotions only where such phrasing could be regarded as a rough, makeshift description of the *behaviour* of individuals (cf. footnote, p. 124) and I hesitate to launch into phrasings which would render my "behaviouristic" references to emotion ambiguous by implying that the behaviour may be the reverse of that primarily appropriate to the emotion.

I grant in general that theories of inversion, etc., are an attempt to express an important truth, but I suspect that the theories cannot be properly formulated or proven till we have techniques for analytical description of gestures, posture, etc., and can define the differences between inverted and direct behaviour.

It is likely that the further apart the personalities evolve and the more specialised they become, the more difficult it will be for them each to see the other's point of view. Finally, a point is reached where the reactions of each party are no longer a striving after the answer which was formerly satisfactory, but are simply the expression of distaste for the type of emotional adjustment into which the other party has been forced. The personalities thus become mutually contra-suggestible. In place of patterns of behaviour which were perhaps originally adopted in an attempt to fit in with the other party, we now have patterns of behaviour which are definitely a reaction against the other party. Thus the schismogenesis takes a new form, and the relationship becomes less and less stable.

If we see schismogenesis as a process in which each party reacts to the reactions of the other, then it follows that the differentiation must, other things being equal, proceed according to some simple mathematical law; even that the specialisation of each party in its particular patterns of behaviour should follow an exponential curve if plotted against time. We cannot, however, hope to get accurate measurements of the progress of the phenomenon in the majority of cases of schismogenesis, but it is possible that in some relatively isolated case some detail of behaviour will serve as an index of degree of specialisation, and we shall be able to judge what sort of curve occurs in that case.

In general, however, not only is the specialisation not susceptible of measurement, but also the progress of the schismogenesis is likely to be modified by the most various factors. We may mention here two types of factors which may be supposed to precipitate and hasten the process:

(a) An individual living in a culture is trained to regard certain patterns of behaviour as commendable, and certain others as wrong. If, in a schismogenesis, he is led to overstep the limits of what is culturally approved, such of his behaviour as passes these limits is likely to have disproportionate effects, not only in making him conscious of the distortion of his personality, but also in driving the other party to over-drastic replies. But against this, it must be remembered that the cultural taboos upon certain sorts of behaviour may be active in restraining the schismogenesis.

(*b*) A second factor which may well hasten the process is a tendency to believe that its outcome will inevitably be tragic. The schizoid individual's preoccupation with destiny may well be not only a realisation that natural process has got him in its clutches; it may be a factor which hastens the process towards its final outcome.

The picture of schismogenesis which I have so far given is of a process inevitably advancing towards such differentiation that some outside factor is bound to precipitate the final collapse. I have described the schizoid as if his degeneration into schizophrenia was inevitable. I have described the breakdown of marriage in such terms that it is inexplicable that any marriage should ever be a success. I have described the difficulties of culture contact in terms which would imply that the final equilibrium which I mentioned could never be achieved. But we know that, in actual practice, schizoid individuals may maintain the same degree of maladjustment for long periods. We know that marriages are sometimes successful, and that they sometimes recover from an unstable condition. We must therefore think of schismogenesis, not as a process which goes inevitably forward, but rather as a process of change which is in some cases either controlled or continually counteracted by inverse processes.

The term "dynamic equilibrium" which I have applied to these balanced states is borrowed from the jargon of chemistry: it is there used to describe apparently stationary equilibria which theoretically can only be described in terms of two or more opposed chemical reactions occurring simultaneously. It is in a sense closely comparable with this that I use the term in the present context. The analogy between the dynamic equilibria which occur in human society and those of chemistry cannot, of course, be pushed very far; but it should serve a useful purpose in reminding us that the position of equilibrium may vary enormously from one schismogenesis to another and from one culture to another. A condition of mutual opposition which would spell disaster for many English marriages may well be an equilibrium position for marriages in other cultural settings; and the equilibrium reached in the sex differentiation among the Iatmul, with its steep ethological contrasts, is very different from the corresponding differentiation in England. The rivalry between initiatory moieties in Iatmul society would seem to be stable, but the state of affairs which obtains between these moieties is one in which not only ritual combat but also continual brawls are culturally normal.

We must expect also to find a very different strength of restraining factors in different contexts of the same culture. I have mentioned above that the complementary patterns which are culturally normal in a relationship between mother and son may become, in our culture, the basis of serious schismogenesis when they are carried over into the relationship between man and wife. We must ask why these patterns are dangerous in one context and safe in another; and it is probable that we must look for the answer either in the presence of adequate restraining factors in the one case and their absence in the other, or in the fact that these complementary behaviour patterns are culturally expected and therefore easily tolerated in one relationship, while they are in some degree culturally disparaged in another.

We have as yet no carefully collected material which will demonstrate the existence of such processes, but from our general knowledge of human behaviour, it is possible to make a number of suggestions as to the nature of the factors which restrain schismogenesis. These factors may be classified into two main groups. We began this chapter with a consideration of the processes which, by moulding each individual to his appropriate ethos, tend to maintain the *status quo*. From this we went on to consider schismogenesis as a process which, if it were permitted to run its course, would bring about changes in the *status quo*. Here again, in considering the factors which control schismogenesis, we have to deal with some whose only effect is to limit the progress of the schismogenesis, and with others which, if they were permitted to run their course, would bring about changes in the *status quo*.

In the first category, the factors that we have to consider are processes of the moulding of the individuals. While the

Iatmul boy is trained to admire harshness and himself to be harsh, he is also trained to regard certain extremes of harsh behaviour as reprehensible. More especially he learns that such extremes are dangerous, and that others would take vengeance upon him if he put them into practice. There is thus an upper limit of tolerance of assertive behaviour, and it is likely that a similar limit of tolerance exists in the case of exhibitionism.

In my account of Iatmul ethos, I have said little of its upper limits. To a European, the emphases of Iatmul masculine ethos are so striking that he does not easily realise that the ethos is not only promoted to the point of harshness which he observes, but is also prevented from developing beyond that point. There are, however, indications in the material which I collected which show the reality of this upper limit of tolerance. I have mentioned that the individual whose photograph (Plate XXII) gives the best impression of Iatmul masculine ethos was actually rather too unstable for native taste. Another example, which illustrates the position of the man who carries Iatmul ethos too far, may be seen in the story of Mwaim-nanggur (*Oceania*, 1932, Plate VI). This hero had a beautiful nose, and he originated the process of extracting sago; but all the women loved him, and, with the detached humour typical of the Iatmul, he arranged them in two rows, those with hair on their pubes on one side and those without on the other: he then copulated with them all. At this point, the limiting mechanisms came into play, and Mwaim-nanggur was murdered by the husbands of the women.

Even success in head-hunting may be carried too far, and it is still remembered against Malenembuk, who died about four generations ago in Palimbai, that when he was on the prow of the war canoe he always killed and left no one for others to kill.

In general, while the Iatmul culture emphasises pride and harshness, channels are provided for the legitimate expression of this ethos, and if a man specialises too far in violence, his wives will run away from his house, his brothers-in-law will turn against him, and he will live under a threat of perhaps violent death, but certainly sorcery.

Among the processes which, like schismogenesis, would be progressive in their unrestrained action, but which may be balanced against the latter process, we may count the following:

1. It is possible that actually no healthy equilibrated relationship between two groups or two individuals is either purely symmetrical or purely complementary, but that every such relationship contains elements of the other type. It is certainly easy to classify relationships into one or the other category according to their predominant emphasis, but it is possible that a very small admixture of complementary behaviour in a symmetrical relationship, or a very small admixture of symmetrical behaviour in a complementary relationship, may go a long way towards stabilising the position. For example, the Squire is in a predominantly complementary and not always comfortable relationship with his villagers, but if he participates in village cricket (symmetrical rivalry) but once a year, this may have a curiously disproportionate effect in easing the schismogenic strain in the relationship.

In Iatmul culture, we have noted a number of small inconsistencies in the ethos of each sex. In the case of the men, we noted that two types are singled out for approbation: first and foremost the man of violence, but after him the man of discretion. It is possible that the trust and liking which other individuals have for the discreet man may well contribute an undercurrent of almost Iatmul-womanly ethos, in terms of which some of the patterns of behaviour between the sexes may be symmetrical. Among the women, we have seen even more clearly a double emphasis running through their ethos; while in their everyday life they are jolly and co-operative, upon ceremonial occasions they tend towards the proud ethos of the men. It is certain that they are then admired by the men, and these occasions may well be of considerable importance in reducing the opposition between the sexes.

2. It is possible that where the principal emphasis in a schismogenic relationship is upon one complementary pair of patterns, that the admixture of complementary patterns

from another potentially schismogenic pair may stop the first schismogenesis. To give a hypothetical case, it is likely that, in a marriage in which there is a complementary schismogenesis based upon assertion-submission, illness or accident may shift the contrast to one based on fostering and feebleness. Such a shift may bring instant relief to the schismogenic strain, even though the schismogenesis had previously reached the limits of cultural tolerance.

3. Similarly, a sudden change in the terms of a symmetrical rivalry may relieve the schismogenic strain. An example (either of this or of No. 2) may be seen in Iatmul culture in what I have termed the diversion of harshness into buffoonery. Here we have a sudden shift in behaviour patterns, prompted perhaps by an individual's consciousness of strain. It is likely that this shift immediately retards or reverses the former schismogenesis.

4. It is reasonably certain that either a complementary or a symmetrical schismogenesis between two groups can be checked by factors which unite the two groups either in loyalty or opposition to some outside element. Such an outside element may be either a symbolic individual, an enemy people, or some quite impersonal circumstance which gives satisfaction or misery to both groups alike. Examples of this process are common enough: the celebration of a Jubilee is useful in reducing the schismogenic antipathy between the different strata of society; and an outside war is well known as a device for reviving a nationalism which suffers from schismogenic splits. In Iatmul culture the same mechanism presumably occurs. It is probable that head-hunting warfare against outside communities contributed in no small measure to control the schismogeneses between the sexes and between the initiatory moieties; and it is possible that nowadays the complementary relationship between natives and Europeans has to some extent taken over analogous functions.

But we may note that wherever the outside element is a person or a community, the relationship between this outside group and the original two groups is itself always schismogenic, and that this second schismogenesis may be even more difficult to control than the first.

5. The process mentioned under No. 4 presents a number of problems, the answers to which may perhaps reveal other mechanisms which control schismogenesis. These are the problems of the effect of an external schismogenesis upon the patterns of behaviour *within* each of the groups involved in it. The problem may be stated in almost mathematical form: if group *A* has a complementary schismogenic relationship with group *B*, what relation can we observe between the patterns of behaviour which the members of *A* exhibit in their dealings with *B* and the patterns of behaviour which they exhibit in their dealings with one another? Similarly, what are the patterns of behaviour inside group *B* among individuals who, in their dealings with *A*, exhibit patterns complementary to those exhibited by *A*? We should expect that, if the schismogenic relation between *A* and *B* is permitted to proceed to a point at which marked distortion occurs in the personalities concerned, their behaviour inside their respective groups should be markedly affected by this distortion; and since the direction of the distortion is different in the two groups, we should find that in each group there is developed a special ethos, related in some simple way to the terms of the schismogenic contrast.

When the problem is stated in this way, it involves two subsidiary problems. First, that of the mechanism of equilibration of hierarchies. In a hierarchic series, we may see each grade as involved in a complementary schismogenic relationship with the grade above, and in another similar relationship with the grade below. We may suppose that the distortion of personality caused by these schismogeneses is to some extent relieved by the fact that in the middle groups of the series, each individual is prompted to exhibit one set of patterns of behaviour to members of the higher groups and a set of complementary patterns to members of the lower groups. But this compensation is not permitted to the highest and the lowest members of the series, and we must expect to see signs of unbalanced schismogenesis in these individuals.

The second problem is closely related to that of the hierarchies. If the behaviour between individuals within each of groups *A* and *B* is a function of the behaviour which each group exhibits in its complementary relationship with the other, then what are we to say of schismogeneses which occur, not between groups of individuals, but between isolated pairs of individuals? Are we to suppose that compensations occur within the personality of each? And is it possible to hope that, when we know how the behaviour between individuals is influenced by the schismogenesis in which their group is involved, an understanding of this process will give us a jargon for describing the compensations which occur *within* the individual?[1]

6. Two special cases of the control of schismogenesis by outside stimuli may be mentioned: (*a*) the possibility that if individuals could become conscious of the schismogenic process in which they are involved, they might react to this rather than continue to specialise in their schismogenic patterns; (*b*) in Iatmul culture, it is probable, not only that the schismogenic relation between the men and the women contributes to controlling the symmetrical schismogenesis between the initiatory moieties, but also that the latter schismogenesis is to some extent controlled by the orientation of the men's attention towards the secrets of initiation. We may see here a complicated example involving a whole series of mechanisms of the types referred to under Nos. 4 and 5.

7. It is possible that in a complementary schismogenesis the patterns of behaviour between the two groups concerned may be such that, while each group diverges farther and farther from the other, the members of each also become more and more dependent upon the complementary behaviour of the members of the other group, so that at some point in the progress of the schismogenesis a balance will be reached

---

[1] In the case of schismogenesis between groups, we noted that a schismogenic relationship with an outside enemy might *prevent* schismogenesis between the classes within a nation, but we also noted that outside admiration might *induce* a schismogenic split within the personality of an individual. This is probably not a discrepancy between individual and group phenomena but an indication that an external schismogenesis will promote an internal only if both splits are in the same terms. A war between two capitalist states may reduce the schismogenesis between Fascists and Communists in each; but a war with Russia, especially if phrased as a "war against Communism", might have very different effects.

when the forces of mutual dependence are equal to the schismogenic tendency.

8. Lastly, it is certain that a schismogenesis, besides being controlled by other schismogeneses either inside or outside the groups concerned, may also be counteracted by inverse progressive changes in the relationship between the groups concerned. These processes are, like schismogeneses, cumulative results of each individual's reactions to the reactions of members of the other group, but the inverse process differs from schismogenesis in the direction of the change. Instead of leading to an increase in mutual hostility, the inverse process leads rather in the direction of mutual love.

This process occurs not only between groups but also between pairs of individuals; and on theoretical grounds, we must expect that if the course of true love ever ran smooth, it would follow an exponential curve.

CHAPTER XIV

The Expression of Ethos in Naven

SEX ETHOS AND *NAVEN*

THE most important generalisation which can be drawn from the study of Iatmul ethos is that in this society each sex has its own *consistent* ethos which contrasts with that of the opposite sex. Among the men, whether they are sitting and talking in the ceremonial house, initiating a novice, or building a house—whatever the occasion—there is the same emphasis and value set upon pride, self-assertion, harshness and spectacular display. This emphasis leads again and again to over-emphasis; the tendency to histrionic behaviour continually diverts the harshness into irony, which in its turn degenerates into buffooning. But though the behaviour may vary, the underlying emotional pattern is uniform.

Among the women we have found a different and rather less consistent ethos. Their life is concerned primarily with the necessary routines of food-getting and child-rearing, and their attitudes are informed, not by pride, but rather by a sense of "reality". They are readily co-operative, and their emotional reactions are not jerky and spectacular, but easy and "natural". On special occasions, it seems, the women exhibit an ethos modelled upon that of the men, and it would appear from our consideration of Preferred Types that certain women are admired for what we may describe as Iatmul-masculine characteristics.

If we return at this point to the problems presented by the *naven* ceremonies, we see these problems in a new light. The elements of exaggeration in the *wau*'s behaviour appear, not as isolated oddities, but as patterns of behaviour which are normal and ordinary in Iatmul men. This answer may seem rather uninteresting but it involves a major generalisation about cultural behaviour, and in science every step is a demonstration of consistency within a given sphere of relevance. We might perhaps have studied this consistency more fully but the answer would still have been of the same type. To pursue the matter further we should be compelled to shift to some other scientific discipline, e.g. to the study of character formation.

In the case of the women, with a double emphasis running through their ethos, their *naven* behaviour can be completely classified as consistent either with their everyday ethos or with their special occasional pride. All the behaviour of the mother is patterned upon submission and negative self-feeling. Her action in lying naked with the other women while her son steps over them, and the cliché "that so small place out of which this big man came", are perfectly in keeping with the everyday ethos of Iatmul women, and constitute a very simple expression of her vicarious pride in her son. Thus the problem of the mother's behaviour, like that of the *wau*'s exaggerations, may now be referred to other scientific disciplines.

In the behaviour of the transvestite women, the father's sister and the elder brother's wife, we may see an expression of the occasional pride such as women exhibit on the rare occasions when they perform publicly with men as an audience.

Examination of Iatmul ethos has accounted for the tone of behaviour of the various relatives in *naven*, but there are many details which cannot be thus summarily dismissed. Consider the *wau*: his buffooning is normal, but that is no reason why he should dress as a woman in order to be a buffoon, and, as we have seen above, the structural premises within the culture, whereby the *wau* might regard himself as the *laua*'s wife, are still not a dynamic factor which would compel either the community or the *wau* to emphasise this aspect of the *wau-laua* relationship. We have still to find some component of the *naven* situation which shall act in a dynamic way to induce transvesticism.

I believe that we may find an answer to this problem if we examine the incidence of transvesticism in European society. In the *naven*, the phenomenon is not due to abnormal hormones nor yet to the psychological or cultural maladjustment of the transvestites; and therefore in looking for analogous phenomena in Europe we may ignore the aberrant cases and should examine rather the contexts in which some degree of transvesticism is culturally normal.

Let us consider the case of the fashionable horsewoman. Her breeches we may perhaps regard as a special adaptation, and she will say that her bowler hat is specially designed to protect her head from overhanging trees: but what of her coat, tailored on decidedly masculine lines? She wears feminine evening dress at the hunt ball, and her everyday behaviour is that of a culturally normal woman, so that we cannot explain her transvesticism by a reference to her glands or abnormal psychology.

The facts of the matter are clear: a culturally and physically normal woman wears, in order to ride a horse, a costume unusual for her sex and patterned on that of the opposite sex; and the conclusion from these facts is equally obvious: since the woman is normal, the unusual element must be introduced by the act of riding a horse. In one sense, of course, there is nothing exceptional in a woman's riding— women have ridden horses for hundreds of years in the history of our culture. But if we compare the activity of riding a horse with other activities which our culture has decreed suitable and proper for women, we see at once that horse-riding, which demands violent activity and gives a great sense of physical mastery,[1] contrasts sharply with the great majority of situations in a woman's life.

The ethos of women in our culture has been built up around certain types of situation and that of men around very different situations. The result is that women, placed by culture in a situation which is unusual for them but which is usual for men, have contrived a transvestite costume, and this costume has been accepted by the community as appropriate to these abnormal situations.

With this hint of the sort of situation in which trans-

---

[1] In Freudian phrasing the act of riding a horse might be regarded as sexually symbolic. The difference between the point of view which I advocate and that of the Freudians is essentially this: that I regard such sexual symbols as noses, flutes, *wagan*, etc., as symbolic of sex ethos, and I would even see in the sexual act one more context in which this ethos is expressed.

vesticism may be developed, we may return to Iatmul culture. First let us consider the contexts in which partial transvesticism occurs, namely, in the case of women who take part in spectacular ceremonies (cf. Plate XIX). Their position is very closely analogous to that of the horsewoman. The normal life of Iatmul women is quiet and unostentatious, while that of the men is noisy and ostentatious. When women take part in spectacular ceremonial they are doing something which is foreign to the norms of their own existence, but which is normal for men—and so we find them adopting for these special occasions bits of the culture of the men, holding themselves like men[1] and wearing ornaments which are normally only worn by men.

Looking at the *naven* ceremonies in the light of this theory we can recognise in the *naven* situation conditions which might influence either sex towards transvesticism. The situation may be summed up by saying that a child has accomplished some notable feat and its relatives are to express, in a public manner, their joy in this event. This situation is one which is foreign to the normal settings of the life of either sex. The men by their unreal spectacular life are perfectly habituated to the "ordeal" of public performance. But they are not accustomed to the free expression of vicarious personal emotion. Anger and scorn they can express with a good deal of over-compensation, and joy and sorrow they can express when it is their own pride which is enhanced or abased; but to express joy in the achievements of another is outside the norms of their behaviour.

In the case of the women the position is reversed. Their co-operative life has made them capable of the easy expression of unselfish joy and sorrow, but it has not taught them to assume a public spectacular role.

Thus the *naven* situation contains two components, the

---

[1] In theatrical representations, humorous journalism and the like, there is a common belief that the postures, gestures, tones of the voice, etc., of the horsewoman are to some extent modelled on those of men, and we might see in this an analogy with the proud gestures of the transvestite and semi-transvestite Iatmul women. I am uncertain, however, to what extent these postures, etc., of the horsewoman occur in real life, and whether they are not perhaps imaginary.

element of public display and the element of vicarious personal emotion; and each sex, when it is placed by culture in this situation, is faced by one component which is easily acceptable, while the other component is embarrassing and smacks rather of situations normal to the life of the opposite sex. This embarrassment we may, I think, regard as a dynamic[1] force which pushes the individual towards transvesticism—and to a transvesticism which the community has been able to accept and which in course of time has become a cultural norm.

Thus the contrasting ethos of the two sexes may be supposed to play and to have played in the past a very real part in the shaping of the *naven* ceremonies. It has provided the little push which has led the culture to follow its structural premises to the extremes which I have described. When the women take part in spectacular ceremonial other than *naven*, the structural premises which might justify complete transvesticism are lacking, and the women content themselves with wearing only a few masculine ornaments.

Lastly we may consider the adoption of widow's weeds by the *wau* and the wearing by the women of the best masculine ornaments obtainable. The former is no doubt a buffooning expression of the men's distaste for the women's ethos. We have seen that the context of mourning is one in which the differing ethos of the two sexes contrasts most strongly and most uncomfortably, and the wearing of a widow's weeds by the *wau* is clearly on all fours with the men's trick of caricaturing the dirging of the solitary widow as she paddles her canoe to her garden. In shaming himself he is, incidentally, expressing his contempt for the whole ethos of those who express grief so easily.

The women on the other hand have no discernible contempt for the proud male ethos. It is the ethos appropriate to spectacular display, and in the *naven* they adopt as much of that ethos as possible—and even exaggerate it, gaily scraping the lime sticks in their husbands' gourds till the serrations are quite worn away. Here it would seem that

---

[1] Cf. footnote, p. 121.

their joy in wearing masculine ornaments and carrying on in
the swaggering ways of men has somewhat distracted them
from the business in hand—that of celebrating the achievement of a small child. Apart from the one incident in which
the women lie down naked while the hero steps over them,
the *naven* behaviour of the women is actually as irrelevant
as that of the men. Thus the presence of contrasting ethos
in the two sexes has almost completely diverted the
naven ceremonial from simple reference to its ostensible
object.

Nevertheless, since *naven* behaviour is the conventional
way in which a *wau* congratulates his *laua* upon any achievement, there is no doubt that this behaviour, distorted and
irrelevant as it may seem to us, is yet understood by the *laua*
as a form of congratulation.

KINSHIP MOTIVATION AND *NAVEN*

I have shown above how the *naven* ceremonial is an expression of sex ethos, that is to say, I have shown the motives
which we may attribute to the relatives concerned in *naven*
qua members of one or the other sex. And this attribution
of motive was an extrapolation from general conclusions
which we drew from the ethos of Iatmul culture. We have
now to consider the motivation of the various relatives *qua*
men and women in particular kinship positions. Here again
we shall use our general ethological conclusions to guide us
in selecting the motives which we attribute to the various
relatives. But instead of dealing with the relatives in groups
according to their sex, we must examine the kinship motivation of each type of relative separately:

wau (mother's brother). In the case of the *wau*, the
symbolism of his behaviour is in a structural sense clearly
interpretable in terms of his position as "mother" and "wife"
of the *laua*. But symbolism has always more than a structural side, and we should see the symbolic acts—the rubbing
of the male *laua*'s shin with the buttocks, and the giving

birth to the female *laua*—as acts informed with emotion and inarticulate[1] reference, as well as with "logic". We have therefore to consider what motivation, conscious or otherwise, real or culturally presumed, leads the *wau* to stress these particular aspects of his relationship with his *laua*.

The undoubted fact that any emotion and inarticulate motivation which we may impute to the classificatory *wau* will be factitious need not cause us any dismay; indeed, it would be difficult to explain any ritual without invoking as its mainstay the very real force of trumped-up or culturally presumed emotion.

In the chapter dealing with the sociology of *naven* we have seen that the classificatory *wau* is attempting to bolster up the allegiance between himself and his *laua*. He is trying to bring this classificatory tie into such prominence that it shall have the binding power of a tie between a boy and his own mother's brother—perhaps even the binding power of the tie with own mother. Symbolically, he does this by acting out two sides, the maternal and the wifely, of the own mother's brother's relationship with the *laua*. But it is questionable to what extent real wifely or maternal emotion can be attributed even to the own *wau*; so that the performance of the classificatory *wau* contains, in a sense, two degrees of falsity. He is acting the part of an own *wau* who is acting the parts of mother and wife.

In spite of this, it is still worth while to consider the motivation of the performances. The fact that the behaviour is acted rather than spontaneous will not, of course, render the motivations which it portrays any the less unconscious. An actor playing *Hamlet* may behave as if he himself were driven by Hamlet's emotions, but the actor may remain to a great extent inarticulate as to the nature of the drives which he is expressing.

---

[1] In the present context, I have deliberately written "inarticulate" in preference to the more fashionable "unconscious". It is to me not perfectly clear that there is any difference in meaning between these two terms, but the use of the second would imply adherence to a theory of the stratification of mind with the validity of which we are not here concerned.

We have to consider, then, the psychology of two[1] persons:

1. The real motivation of the classificatory *wau*. Under this head will fall the series of emotional factors which have already been sufficiently indicated: his pleasure in his own buffooning performance; his scorn of the female ethos; his awkwardness in expressing vicarious pride; his desire for the allegiance of the *laua*, etc.

2. The inarticulately presumed motivations of the own *wau*. Under this head we have to consider possible drives which might lead a *wau* to stress the maternal and wifely sides of his relationship with his *laua*.

Here we are on very treacherous ground, since it is singularly easy to construct hypotheses about inarticulate motivation, but exceedingly difficult to test them, since no two such hypotheses are ever mutually exclusive, and the psychoanalysts will tell us to regard every native statement as meaning either what it says, or the opposite. I shall, however, put forward a series of possible hypotheses in order to indicate what sort of answer may be expected in the present case, but I would not stress any one hypothesis as more probable than others which might be constructed.

In this connection there are three myths which are probably relevant. These are all stories in which a *wau* kills his *laua*. In each case the murder is followed by a more or less complete disintegration of the community, and the myth becomes a history of colonisation—of the scattering of groups from the disintegrated centre.

These myths may be (1) a reference to the importance of matrilineal ties in the integration of the community, or (2) the expression of an underlying opposition or slight hostility between *wau* and *laua*. These interpretations are not mutually exclusive. Rather, I believe that we should see the two interpretations as mutually supplementary. On the one hand, the sociological importance of the tie may well lead to an increased uneasiness about the opposition which is inherent in it, while on the other hand, if there were no opposition between *wau* and *laua*, there would be no need to stress the sociological

---

[1] In the rarer cases in which own *wau* performs *naven*, the matter is, of course, simpler, and the second set of emotions proportionately more real.

importance of allegiance between them. In general, the myths may be taken as an indication of some opposition between *wau* and *laua* and some uneasiness about that opposition. In fact, it would appear that the *wau-laua* relationship is ambivalent.

It is in such ambivalent feelings that we may expect to find the motives for the *wau*'s symbolic behaviour and we must therefore examine the *wau-laua* relationship to find what aspects of it may be regarded as the cause of these mixed feelings.

We have not far to seek for the basis of the *wau*'s friendly feelings for his *laua*, since the two are united by ties of allegiance and common interest. Following the lines of our structural analysis, we may attribute the *wau*'s friendly feelings to any and all of the four identifications which define his position. But the basis for the trace of hostility is more obscure. In our structural analysis we found that the *wau-laua* relationship contains little or nothing which might be regarded as *sui generis*, and that all its details could be classified satisfactorily as due to the various identifications of the *wau* as (*a*) mother, (*b*) wife, (*c*) wife's brother, and (*d*) father of his *laua*.

It is therefore reasonable in our search for the basis of the *wau*'s hostility to suppose that this arises from one or other of these aspects of the relationship.[1] Of these aspects, the last, that due to the *wau*'s identification with the father, might perhaps give a possible phrasing of the *wau*'s hostility as an expression of the ambivalent situation between father and son. But I think that this solution must be dismissed, for two reasons:

1. This solution is incompatible with the fact that, not only the *wau*, but also the wife's brother sometimes rubs his buttocks on his sister's husband's leg, when the latter marries his sister.

2. The identification of *wau* with father is only a very minor motif in the patterning of the *wau-laua* relationship.

[1] It is, of course, possible that the *wau*'s "hostility" may be due in part to *interaction* between two or more of these aspects of his position. This complicated type of explanation I have ignored in favour of the simpler solution based upon the brother-in-law relationship.

Of the three remaining aspects of the *wau-laua* relationship, two, the maternal and the wifely, may likewise be excluded on various grounds:

1. These are the aspects which are actually stressed in the *naven* and, though it is conceivably possible that the *wau* might stress those very aspects which were the cause of his trace of hostility, it is much more likely that he gets relief from his uneasiness by stressing the other aspects.

2. It is exceedingly difficult and perhaps impossible, if we attribute the basis of the opposition to either of these aspects, to construct any theory which will cover the *wau*'s *naven* behaviour to *laua*s of both sexes, and the *naven* behaviour of the wife's brother.

3. The mother's relationship to her child is not ambivalent.

We are therefore driven to conclude that the *wau*'s trace of hostility is due to his position as a brother-in-law of his *laua*. And on this theory we may see, in his stressing of the "mother" and "wife" relationships, a denial of his position as wife's brother.

This theory receives support from examination of the brother-in-law relationship, since we can clearly detect elements of opposition[1] in that relationship (cf. p. 78). But the position is still not perfectly clear. The familiar processes of compensation and displacement cannot be appealed to in the present case. Typically, the *wau* having ambivalent feelings towards his sister's husband might either exaggerate his friendliness in his dealings with that relative, or he might vent his hostility upon the *laua* who would then be a substitute for the sister's husband. In practice, he does neither of these things, but exaggerates his friendliness towards his *laua* rather than his hostility. Further, there is no reason why the hostility felt for the sister's husband should be displaced on to the *laua*. We have seen that in general this

[1] It should be possible to demonstrate that *tawontu*'s (wife's brother's) hostility is consistent with the male ethos of the culture, and I think that the clue to this problem perhaps lies in a pride which the men occasionally exhibit, a pride in the possession of women-folk. Cf. also the cliché "She is a fine woman" as a justification for endogamy (p. 91). Unfortunately I have not the detailed material necessary for a comparison of attitudes of men towards their sisters with the general male ethos.

hostility is openly handled and acknowledged. The brothers-in-law may either joke about their mutual obligations; or, if the wife's brother's resentment grows strong, he is expected to have recourse to sorcery. It is true that the quarrels of brothers-in-law are phrased in terms of indebtedness and bride price in a context in which a Freudian might suspect sexual jealousy to be the underlying motive, but still the hostile feelings of the wife's brother are in no sense denied expression. There is nothing in all this which would lead him to seek a substitute for his sister's husband.

It is, however, possible to see the *wau* as a man guilty of overt hostility to his sister's marriage, who therefore makes exaggerated amends to her child, and a tenable hypothesis can be constructed upon this basis. But, equally, we may see the *wau*, not as a guilty man making amends, but as an innocent man protesting his innocence. He may have opposed his sister's marriage and have harboured ill-feeling for her husband. In any case it is culturally assumed that some opposition exists between brothers-in-law. But he has no quarrel with the offspring of that marriage, who are indeed connected with him by blood,[1] interest and allegiance. The offspring, however, are also connected no less intimately with the father, and it is this fact which might constitute an imputation that the *wau* is perhaps hostile towards them. It suggests that the *wau*'s hostility, previously centred on the sister's husband, might perhaps spread to include the *laua*. This imputation the *wau* is perhaps denying when he stresses the fact that he is a mother and a wife.

Similar phenomena occur in our own culture, and it is not uncommon to see uncles and aunts who have consistently opposed the parents' marriage later falling over themselves in their efforts to make friends with their nephews and nieces. Their position is analogous to that of the Iatmul *wau*, and the chief difference between them is that the latter falls over, not metaphorically, but really.

Whether we see these agnatic relatives as innocent persons

---

[1] In the native theory of conception, blood and flesh are believed to be products of the mother, while the bones of the child are contributed by the father.

protesting their innocence or as guilty persons making amends, will depend—in the absence of additional facts—upon the view which we take of psycho-analytic theory. We may even, according to our taste, combine the two hypotheses and say that the *wau*, guilty of overt hostility towards his sister's husband, therefore protests his innocence with the greater emphasis when a similar offence is imputed to him by the circumstances of his relationship with his *laua*, about whom he really feels innocent. A whole series of varying phrasings of this sort may be devised, and each of them will perhaps be untrue, because it will err on the side of being too definite; while any attempt to choose between them will be based upon the fallacy that the less articulate levels of the mind are as discriminating and as precise as the scientist.[1]

In any case, we may add to our list of factors affecting the *wau*'s position the opposition which is present in the brother-in-law relationship and its repercussions upon that between *wau* and *laua*, without defining the exact mechanism of these repercussions. Further, it may be noted that these additional factors may be regarded as to some extent dynamic —as pushing the *wau* towards transvesticism along the lines laid down in the structure of the kinship system.

Finally, one other possible motive for the *wau*'s behaviour in *naven* may be mentioned. We have seen in our consideration of the boy's relationship to his father's and his mother's clans that, while in matters of economics the boy is grouped with his father, his feats are regarded as the achievements of his mother's clan (cf. p. 48). In view of these premises, we may suppose that the *naven* behaviour of the *wau* is an act which symbolically lays claim to the *laua*'s achievement, and the value which this culture sets upon pride and achievement no doubt here plays a part. But such a claim can only be vicarious, and here the male sex ethos is again relevant and

---

[1] It is, however, possible that a more detailed study of the standardised attitudes and behaviour of the men, when though innocent they are accused and again when, guilty, they are making amends, might make possible a more definite statement of the *wau*'s motives. To some extent, at least, the vagueness of the statement here offered is the inevitable result of attempting an analysis of motives with an insufficient knowledge of the ethology of the culture.

indicates that in this act the *wau* will be playing the role of a woman. Thus sex ethos and kinship position work together, and each is expressed in the *wau*'s behaviour.

tawontu (wife's brother). From what has been said of the *wau* in *naven*, it might appear that the *tawontu*'s behaviour, in rubbing his buttocks on his *lando*'s shin on the occasion of the latter's marriage, is in some sense the primary form of *naven*. We are, however, not engaged in a study of the history of *naven*, and it is certain that at the present time this particular gesture is regarded as especially characteristic of the *wau*.

The factors which affect the two relatives are alike in many respects, but the behaviour of *tawontu* differs from that of *wau* in certain significant details: (1) The *tawontu* performs only that ritual which stresses his position as "wife" and omits that which would imply that he is a mother—an omission which is no doubt correlated with his structural position. (2) The *tawontu*, so far as I know, only performs his gesture on the occasion of the marriage of his sister. (3) It is, I believe, only the *own* wife's brother who performs in this way, whereas in the case of *wau*, it is usually the classificatory relative who performs *naven*.

The second and third of these points of difference are probably connected with the fact that in general the *tawontu* is not constrained to hide or deny the ambivalence of his feelings. But on the occasion of his sister's marriage—a marriage which he himself has perhaps arranged (cf. p. 79)— it is understandable that he may feel constrained to deny the existence of any trace of hostility[1] towards her husband; and this denial he expresses by emphasising the fact that he is not an affinal relative, but a partner in the marriage, a "wife-man".[2]

nyaï, *nyamun* and *tshuambo* (father, elder brother and younger brother). The only other male relatives whose position we have to consider are the father and brothers of the

---

[1] As in the case of *wau*, various phrasings of the *tawontu*'s motives may be substituted for that here given.
[2] Literal translation of the term *tawontu* (*tagwa-ndo* = wife-man).

hero. All these relatives may take a passive part in *naven*; they may be beaten by their *tshaishi* (elder brothers' wives) and the father may help the hero to find valuables to present to his *wau*. But, apart from being thus secondarily involved, neither father nor brothers take any active part in *naven*.

Since, however, his father and brothers are the boy's nearest male relatives, we might reasonably expect that they too would feel impelled to celebrate his achievements; it therefore becomes necessary to account for their inactivity.

We may first consider the conditions—sociological, economic, structural and emotional—which are probably connected with the father's inactivity:

1. There is no increase in the integration of the community to be gained by a further stressing of the patrilineal links. In such a society as that of the Iatmul, the solidarity within any group depends to a very great extent upon opposition between that group and others outside it: at the same time, it is probable that the greater the solidarity within a group and the greater its strength, the more likely it is to oppose other groups or to ignore its obligations to them. From this it follows that any further strengthening of the patrilineal ties may lead to a weakening of those matrilineal and affinal ties upon which, as we have seen, the integration of the community depends.

2. On the economic side, there is no occasion for presentation of valuables between father and son, because the respective property rights of these two relatives are very much less differentiated than those of *wau* and *laua*. Father and son do not hold their property on a communal basis and there is, as we have seen, a strong feeling against the father accepting food from the son: but there is a general assumption that the son will inherit or is already inheriting from the father, and that the father will do what he can to provide his son with valuables for bride price, etc. Such gifts from father to son are not in any way comparable with the ceremonial presentations[1] which take place between *wau* and *laua*.

---

[1] It would seem that in Iatmul culture ceremonial giving does not occur in purely complementary relationships. It is perhaps a device for stressing the complementary aspects of mixed relationships. (Cf. p. 270.)

Similar considerations apply to the relationship between brothers. They do not hold their property on a communal basis, but the differentiation of their respective rights is still so vague that they are culturally expected to quarrel over their patrimony.

3. There are no elements in the structural position whereby either father or brothers might divert their celebration of the boy's achievement into some form of buffoonery, by acting the part of some other relative. If they were to take any part in the *naven* they would be compelled (like the mother) to stress their *own* relationships with the hero.

4. This last consideration brings us to a very real positive bar on any *naven* activity on the part of the father. His relationship with his son is, as we have seen, ambivalent; but the ambivalence is not of a sort which can be relieved by stressing either of its components. The father is, on the one hand, a disciplinary authority and, on the other, he is a relative who is expected to give way before his son's advancement. The excessive stressing of either of these aspects of the paternal relationship leads at once to the embarrassment of both parties, and we have seen that in daily life father and son pattern their behaviour on a rigid middle course, avoiding even any display of mutual intimacy. In such a situation it is clear that the father could not in the *naven* stress or dramatise either aspect of his relationship with his son.

It would seem that even the passive role played by the father, in being beaten by his *tshaishi*, is to some extent incompatible with his kinship position. I was told (but I did not observe this) that the own father would be only slightly beaten, but the *tshambwi nyai'* (father's younger brother) would be very much more severely beaten.

We may note the contrast between the father's position and that of the *wau*. The latter's kinship position is defined in terms of his various identifications; one of these identifications (*qua* wife's brother) is embarrassing, and he escapes from this by emphasising the others. But the father has a unitary relationship with his son which is not subdivisible in this way. Although his feelings towards his son are culturally presumed to be ambivalent, he cannot dodge the whole issue by stressing some other side of his position, and if he stressed either half of his ambivalent position he would be embarrassed.

In the case of elder and younger brothers, the positive bar is less clear: but the relationship between brothers is certainly felt to be analogous to that between father and son, though the issues between them are less dramatically and less rigidly defined. On the one hand, brothers are expected to be allies as over against outsiders, but between themselves they are culturally expected to quarrel over the patrimony, and the elder brother has some authority over the younger. I was told (but never observed this) that there is some avoidance between brothers: they would not walk about together much; they would go together in order to do something, but would not do so without a definite purpose. Thus their relationship contains the germ of the same type of ambivalence as that which exists between father and son.[1]

Turning now to the *naven* behaviour of the women we find a very considerable difference between their motivation and that of the men. It is possible that this impression is due to the inadequacy of my information, but I believe that the contrast is a real one.

In the case of the *wau* we have found that his behaviour is an expression of the following factors:

1. His structural position in the kinship system and the various identifications which define it.
2. The male sex ethos of Iatmul culture.
3. His need for the allegiance of his *laua*.
4. The ambivalence implied in his kinship position.
5. Economic considerations.

But the women, so far as I know, are not influenced in their *naven* by any motives of the last three types. Rather, I believe that we should see the women's *naven* as almost solely the expression of their ethos and of the structural identifications implied in their kinship status. I am inclined to see the *naven*

[1] As evidence that this relationship is really felt to be analogous to that between father and son cf. the staggering of brothers (p. 246) and the use of the refrain "*nyai' nya! nyamun a!*" in name songs (footnote, p. 39).

of the *wau* as in some sense primary and as seriously motivated by desire for allegiance, economic gain, etc., while
I would see the *naven* of the women as an amusement, a gay
occasion when they embroider upon structural premises
analogous to those followed by the *wau* and when they enjoy
the special privilege of wearing men's attire.

In the case of the *wau*, we observed that only one or two
wau's usually take part, and that these were especially the
classificatory relatives. But there is no such rule in the case
of the women. When the children came back from sago-
making and a *naven* was celebrated for them in Mindimbit,
it appeared as if all the women had gone mad. Every woman
who could regard herself as appropriately related to the
children made a show of *naven* behaviour.

Moreover, in the *naven* for the successful homicide I was
told that *all* the women, with the exception of the hero's
wife and own sister, lie down on the ground for the hero
to stride across them. In this detail we see a pattern of
behaviour, which is perhaps primarily characteristic of the
mother, being adopted by the other women not according
to the details of their kinship status, but simply in virtue of
their sex. While the *naven* of the men is in part controlled
and motivated by advantages which the exploitation of their
kinship position may bring, that of the women is apparently
almost devoid of such motivation.

Correspondingly, on the sociological side I believe that
we should see the function of the women's *naven* as much
more diffuse than that of the *wau*'s *naven*. While the latter
strengthens the solidarity of the community by emphasising
certain specific affinal linkages, the former seems rather to
spread a more diffuse euphoria throughout the community.

One other point which the *naven* behaviour of the women
contrasts with that of men may be mentioned. The *wau*
stresses, in his *naven*, emotions which are freely and easily
expressed by the "mother" whom he imitates, but the women
in their *naven* exaggerate special aspects of the father's and
elder brother's position which these relatives could not freely
express without embarrassment. The father must repress his
disciplinary activities and give way before his son, but the

women have picked upon the disciplinary aspect of father-hood for exaggeration in *naven*. We cannot of course suppose that the repressed emotions of a father are a dynamic force in the shaping of the women's behaviour. But it may be that the women get some extra excitement out of thus exaggerating attitudes which are to some extent culturally taboo in the relatives whom they are aping.

At the beginning of this chapter I have already indicated the ethological basis of the women's behaviour, the enjoy-ment which they get out of their swashbuckling and swagger, and the way in which transvesticism goes hand in hand with public performance. It only remains therefore to consider the symbolism of the various details:

nyame (mother). The behaviour of the mother we have already seen to be a simple straightforward expression of the Iatmul mother's pride in her son. There is in her case no structural basis for transvesticism, and her feelings about her child are not markedly ambivalent; therefore there is in her behaviour none of the elaboration present in that of the other relatives. As to the nakedness of the mother in *naven*, we may compare this with the other contexts in which women are ceremonially naked, namely, during the burial of their nearest male relatives (cf. p. 154), and when they go as suppliants (cf. p. 146). It may seem odd that the mother should adopt exactly the same gesture both to express distress and to rejoice in her son's achievement. But I think that we may regard nakedness in all these three contexts as an extreme expression of abnegation or negative self-feeling, an expres-sion which may be accompanied either by joy or sorrow.

iau (father's sister). The *naven* behaviour of the *iau* is based upon her identification with her brother, the father of the hero. In this identification there are clearly two possi-bilities which might be stressed. The relationship of a father to his son is, as we have seen, ambivalent; the father being impelled both to assert his authority and to give way before his son's advancement. Either of these aspects of the ambi-valent relationship might have been adopted by the *iau*, and she has actually chosen the assertive side of fatherhood; she

puts on the finest of homicidal ornaments and beats her
"son". The alternative behaviour would have been to dramatise the father's retirement at his son's achievement. We may
suppose that such a role would be to a Iatmul woman rather
uninteresting and inappropriate to her transvesticism. She
would be adopting male costume merely to play a role more
appropriate to female ethos.

There is so far as I know no detail of dramatisation
referring to the possibility that the hero may marry *iau*'s
daughter; and the absence of such a feature may be noted as
additional evidence for supposing that this marriage is a
recent innovation (see footnote, p. 90).

tshaishi (elder brother's wife). The behaviour of *tshaishi*
is based upon her identification with her husband, the hero's
elder brother; and to a very great extent her behaviour and
kinship position are comparable with those of the *iau*. We
have already seen that an elder brother is in a position
analogous to that of father, so that *tshaishi* identifying herself
with the elder brother is in a position analogous to that of *iau*
who identifies herself with the father.

The only special feature in the *tshaishi*'s position may be
correlated with the levirate whereby it is likely that she will
one day be the hero's wife. This possibility no doubt lends
a certain piquancy to the *tshaishi*'s *naven* and is exploited
to the full in the incident in which the hero spears a fish trap,
the symbol of his *tshaishi*'s womb. The same aspect of the
relationship is again embroidered upon in the behaviour of
the hero's sister.

nyanggai (sister). This relative plays but little part in the
naven, but she is identified with the hero on the occasion
when both step across the prostrate women. Then the sister
expresses with exaggerated dramatisation what we (or she)
may suppose to be the repressed desires of the hero. She
attacks the genitals of the women with her hands and especially
those of *tshaishi*. Upon these she pounces with the exclamation
"A vulva!" but the transvestite *tshaishi* replies "No! A
penis!" This disagreement between them is, no doubt, due
to the fact that the sister regards herself as the potential

husband of *tshaishi* while the *tshaishi* regards herself as potentially an elder brother of the hero.

mbora (mother's brother's wife). This relative identifies herself with the *wau*. Like him she dances with the captured head (cf. Plate V A) and receives presentations of valuables from the *laua*. Inasmuch as she is identified with a man who is himself transvestite, it is not surprising to find that the *mbora* seems to be in some doubt about her sex, so that she appears in *naven* dressed sometimes as a man and sometimes as a woman. Finally, the confusion on this subject would seem to be resolved when she, acting as a man, goes through a ritual copulation with her transvestite husband.

One incident remains to be considered, that in which the *iaus*, in male costume and carrying a feather headdress, come striding across the prostrate *mboras* (also in male costume). Of this I can give only a tentative explanation. We have seen that pride in a man's achievements is regarded as the preroga tive not of his father's clan, to which he himself belongs, but rather of his mother's clan. It would seem that in this piece of fooling the feather headdress is a homicidal ornament.[1] The *iaus* (members of the hero's father's clan) come, as it were, boasting of the achievement. They stride across the *mboras*, who lie on the ground, as the mother lies before her triumphant son. Both their posture and the fact that *mbora* is identified with *wau* would seem to indicate that the *mboras* are representatives of the maternal clan. They jump to their feet and snatch the feather ornament, the symbol of achieve ment, symbolically claiming it as the triumph of the maternal clan.

A piece of supporting evidence for this interpretation may be found in the fact that homicidal ornaments are usually presented by his *lanoa nampa* (husband people) to the killer. This would be consistent with the interpretation which I have put forward.

[1] Unfortunately my informant did not give me the native term for the particular ornament used in this context. He only stated that feathers were stuck into pith, the usual technique in making certain sorts of homicidal headdress. It is probable, however, that the ornament is a badge of homicide.

CHAPTER XV

The Eidos of Iatmul Culture

FROM the foregoing analysis and from the general descriptions which have been given of Iatmul culture, it is evident that this culture has some general pervasive character other than that which it derives from its characteristic ethos. It is a culture which continually surprises us by the mass of structural detail which it has built up around certain contexts. Most conspicuously we have the fabric of fancy heraldry and totemism built up around the personal names and ancestors: but a similar tendency to what we can only describe as hypertrophy may be recognised in the initiatory system, with its plexus of cross-cutting dual divisions and staggered initiatory grades; and again in the *naven* ceremonial, where we have seen that the culture has proceeded upon simple structural premises to such lengths that the *wau* behaves as in some sense a wife of the *laua*.

The culture has apparently some internal tendency to complexity, some property which drives it on to the fabrication and maintenance of more and more elaborate constructs, and since this tendency has evidently contributed to the shaping of *naven*, it is relevant to examine its nature in detail.

I often wondered about this property of the culture when I was in the field, but never saw any clue to an understanding of it. It was only after, in the writing of this book, I had worked out and defined cultural structure and ethos, that I was able to see an approach to the present problem. I did not in the field pay any special attention to the points which I know now might have thrown light on this problem, but since my approach has been but little attempted by others I will give what material I have in order to illustrate the method.

I defined cultural structure as "a collective term for the coherent 'logical' system which may be constructed by the scientist, fitting together the various premises of the culture", and to the word "logical" I added a footnote to the effect that we must expect to find different systems of logic, different modes of building together of premises, in different cultures.

The first point in this definition which we have to consider is the role of the scientist. I said that cultural structure was a system built up by the scientist, and we may phrase our present problem in these terms: why should the materials of Iatmul culture drive the abstracting scientist to greater complexities of exposition than are demanded by, for example, the culture of the Arapesh?[1] Is this complexity of exposition the reflection of a real complexity in the culture, or is it only an accidental product resulting perhaps from a disparity between the language and culture of the ethnographer and those of the community he is describing?

Here we are on very difficult ground, and the scientist who has set out the exposition is perhaps the man least fitted to judge of the origin of its complexity. I can only say, first, that the Iatmul culture appeared to me complex and rich before I even attempted a structural analysis, and secondly, that the result of analysis has been to cause the culture to appear, to me at least, not more complex, but simpler. I can only trust to these impressions as evidence that the complexity is not entirely a creation of my own methods of thinking.[2]

This complexity, then, exists in some form in the culture itself, and is built up of premises. The word premise was defined as an assumption or implication recognisable in a number of details of cultural behaviour, and I mentioned that though these assumptions are often only expressed in symbolic terms, in other cases they are articulately stated by the natives.

If we accept (*a*) that the complexity of the cultural structure is a reflection of some property of the culture itself, and (*b*) that the elements of this complexity are to some extent

---

[1] Cf. Margaret Mead, *Sex and Temperament.*

[2] To some extent my methods of thinking—structural, ethological and sociological—are followed by the natives themselves. This matter is discussed at the end of this chapter (p. 250).

present as ideas or assumptions in the minds of the natives, then it follows that any pervasive characteristic of the cultural structure can be referred to peculiarities of the Iatmul mind; that, in fact, we are here dealing with the cultural expression of cognitive or intellectual aspects of Iatmul personality.

So stated, the problem appears closely analogous to that of ethos. In each case we have to account for a generalised tendency which shows itself in the most diverse contexts of the culture, and it would seem that the tendency which we are now investigating bears the same relation to cultural structure as that which ethos bears to motivation. We have seen that ethos constitutes a factor in the determination of the needs and desires of the individuals, and this factor varies from culture to culture. Ethos is the system of emotional attitudes which governs what value a community shall set upon the various satisfactions or dissatisfactions which the contexts of life may offer, and we have seen that ethos may satisfactorily be regarded as "the culturally standardised system of organisation of the instincts and emotions of individuals".

In an analogous way, the tendency which we are now examining is a variable factor in our definition of cultural structure, the meaning of the word "logical", the factor which determines and systematises the building together of structural premises.

From the close analogy between the two problems we may guess that the variable factor in cultural structure is some sort of standardisation of the individuals in the community. We defined social structure in terms of ideas and assumptions and "logic", and since these are, in some sense, a product of cognitive process, we may surmise that the characteristics of Iatmul culture which we are now studying are due to *a standardisation of the cognitive aspects of the personality of the individuals.* Such a standardisation and its expression in cultural behaviour I shall refer to as the *eidos*[1] of a culture.

[1] The sense in which I use the word eidos is, of course, distinct from that in which it was used in Greek philosophy, but since the philosophical term has not been adopted in this form into the English language, it seems legitimate to use the word, eidos, in the present sense.

This exposition of eidos and the definition of social structure are,

We may now return to the Iatmul culture and examine its eidos in some detail. Let us consider first the phenomenon which I have variously described as hypertrophy, structural complexity, or multiplicity of premises, in order to see whether this can be re-phrased in terms of the standardisation of cognitive processes of individuals. In the light of the definition of social structure, this question becomes excessively simple. Social structure is built up of "ideas or assumptions", and the culture is characterised by the multiplicity of elements in its structure; that is to say, it contains a very large number of "ideas or assumptions", products of cognitive process. To re-phrase this in terms of the standardisation of individuals, we have only to say that they are so modified by their culture that their output of cognitive products is increased. In fact, that they are stimulated by the culture to a degree of intellectual activity unusual among primitive peoples.

Here I am not referring to any increase in intellectual endowment, which is probably controlled to a great extent by hereditary factors. But it is worth mentioning that among the Iatmul definite efforts are made to increase the memory endowment of individuals, by means of magical techniques. Soon after birth, a male child is made to inhale smoke from a fire which has been bespelled, in order that the boy shall grow up to be erudite in the totemic names of his clan; and later in life a man may be treated with spells which are believed to act on his heart (the seat of memory) giving him facility in the memorising of name-cycles and spells. In a culture in which such techniques have been evolved it is at least likely that there is a factor of selection which favours those of high cognitive capacity. In the present instance, however, I am referring, not to an increase in endowment, but rather to the way in which the cognitive machinery of the individuals is stimulated to activity, instead of being allowed to remain idle.

With this *caveat*—that we are discussing not endowment

I think, clumsy, but a more exact formulation of the matter is impossible until an adequate enquiry has been conducted into the eidoses and ethoses of various cultures.

but stimulation—we may expect to find that cultures differ enormously in the extent to which they promote intellectual activity: though I know of no field work in which any attention has been paid to this phenomenon. Any mathematical evaluation of the habitual activity of the intelligence of individuals is, of course, out of the question. But there are many facts which indicate that a high degree of such activity is common among the Iatmul and is promoted by the culture.

The present state of our knowledge of cognitive aspects of mental process is so unsatisfactory that it is not possible to describe the Iatmul material as systematically as we might desire. We may, however, very roughly classify these processes into processes of recall and processes of thought, both of which seem to have contributed to the maintenance and development of complexity in the Iatmul culture.

Let us first consider the cultural stimulation of memory.[1] We have already seen that vast and detailed erudition is a quality which is cultivated among the Iatmul. This is most dramatically shown in the debating about names and totems, and I have stated that a learned man carries in his head between ten and twenty thousand names. This figure was arrived at by very rough estimation from the number of name songs possessed by each clan, the number of names in each song, and the general ability of such men to quote, in considerable detail, from the name-cycles even of clans other than their own. The figure must therefore be accepted with caution, but it is certain that the erudition of these men is enormous.

Further it would seem that rote memory plays a rather small part in the achievement of these feats of memory. The names which are remembered are almost all of them compounds, each containing from four to six syllables, and they refer to details of esoteric mythology, so that each name has at least a leaven of meaning. The names are arranged in pairs, and the names in any one pair generally resemble each other much as the word Tweedledum resembles the word Tweedle-

---

[1] I have been much influenced in my own thinking about these problems by Professor Bartlett's book *Remembering*, 1932, which I read after my return from New Guinea.

dee—with the notable difference that the altered syllable or syllables generally have some meaning and are connected together by some simple type of association, e.g. either by contrast or by synonymy. A progressive alteration of meaning may run through a series of pairs.

Thus the series of names contain tags of reference which would make it possible for them to be memorised either by processes of imagery, or by word association. I collected a great quantity of these names and noticed again and again that the *order* in which the pairs were given was subject to slight but continual variation. There is a vaguely defined standard order for the recitation of every series of names. But I never heard any criticism of the order in which names were recited. In general, an informant will alter the order of his recitation slightly every time he repeats the series. Occasionally even, the pairing of the names is altered, but changes of this type are definitely regarded as mistakes.

Bartlett[1] has pointed out that one of the most characteristic qualities of rote remembering is the accuracy with which the chronological sequence of events or words can be recalled. So that from the continual alteration of the order in which the names are given we may deduce that the mental process used is not chiefly that of rote memory. Additional evidence for this conclusion may be drawn from the behaviour of informants when they are endeavouring to recall an imperfectly remembered series of names. I do not remember ever to have heard an informant go back, like a European child, to the beginning and repeat the series of names already given, in the hope that the "impetus" of rote repetition would produce a few more names. Usually my informants would sit and think and from time to time produce a name (or more often a pair), often with a query as to whether that name has already been given—as was frequently the case.

Again, when a Iatmul native is asked about some event in the past, he can as a rule give an immediately relevant answer to the question and does not require to describe a whole series of chronologically related events in order to lead

[1] *Remembering*, pp. 203 and 264–266.

up to the event in question. The Iatmul indulge very little in the sort of chronological rigmarole which, as Bartlett has pointed out, is characteristic of those primitive peoples who have specialised in rote remembering.

One detail of the culture is worth mentioning as likely to promote the higher processes rather than rote memory. This concerns the technique of debating. In a typical debate a name or series of names is claimed as totemic property by two conflicting clans. The right to the name can only be demonstrated by knowledge of the esoteric mythology to which the name refers. But if the myth is exposed and becomes publicly known, its value as a means of proving the clan's right to the name will be destroyed. Therefore there ensues a struggle between the two clans, each stating that they themselves know the myth and each trying to find out how much their opponents really know. In this context, the myth is handled by the speakers not as a continuous narrative, but as a series of small details. A speaker will hint at one detail at a time—to prove his own knowledge of the myth— or he will challenge the opposition to produce some one detail. In this way there is, I think, induced a tendency to think of a story, not as a chronological sequence of events, but as a set of details with varying degrees of secrecy surrounding each— an analytic attitude which is almost certainly directly opposed to rote remembering.

But though we may with fair certainty say that rote memory is not the principal process stimulated in Iatmul erudition, it is not possible to say which of the higher processes is chiefly involved. There are, however, several details of the culture which point to visual and kinaesthetic imagery as likely to be of great importance. In debate, objects are continually offered for exhibition. For example, when the totemic ownership of the Sepik River was in dispute, a shell necklace was hung in the centre of the ceremonial house to represent the river. In the debating, clan *A* claimed that the elephant grass which forms a conspicuous and picturesque fringe along the banks of the river was indubitably theirs; and that therefore the river must belong to them. They accordingly produced a beautiful spear decorated with leaves

of the grass and pointed to it, saying "Our Iambwiuishi!!"[1]
Clan *B* on the other hand claimed that the river was their
snake, Kindjin-kamboi, and their protagonist, Mali-kindjin
went off to get brightly coloured leaves to ornament a representation of his snake which adorned one of the gongs in the
ceremonial house. Again in a debate about the Sun, a
number of the participants dressed themselves up to represent characters in a myth of origin of the Sun.

In the technique of debating the speaker uses bundles of
leaves, beating on a table with them to mark the points of
his speech. These leaves are continually used as visible or
tangible emblems of objects and names. A speaker will say,
"This leaf is So-and-so, I am not claiming that name", and
he will throw the leaf across to the opposition. Or he may
say, "This leaf is so-and-so's opinion", and he will throw it
onto the ground with contempt; or he will sweep the ground
with the leaves, brushing away his opponents' rubbishy statements. Similarly a small empty leaf packet is used as an
emblem of some secret, of which the speaker is challenging
the opposition to show a knowledge: he will hold it up asking
them scornfully if they know what is inside it.

The proneness to visual or kinaesthetic thought is shown
too in the continual tendency to diagrammatise social organisation. In almost every ceremony, the participants are arranged
in groups so that the total pattern is a diagram of the social
system. In the ceremonial house the clans and moieties are
normally allotted seats according to the totemic system of
groupings: but when initiation ceremonies are to be performed this arrangement is discarded and in its place comes
another based upon the cross-cutting initiatory moieties and
grades.[2]

[1] The totemic name of the grass.
[2] For other material of this kind, see *Oceania*, 1932, *passim*; the statement on p. 256 that the village is longitudinally divided between the two
totemic moieties is, however, false. It was collected while I was on a
short visit to a village in which initiation was taking place. We were in
the initiatory ceremonial house, and it would appear that my informant
was so influenced by the initiatory patterning inside the ceremonial house
that he spoke as if that patterning extended to the whole village. As a
matter of fact, the subdivision of the village does on the whole follow
the moiety divisions, but is *transverse* and not *longitudinal*. The village
of Mindimbit in which I chiefly worked on the first expedition was
irregularly laid out owing to the swampiness of the ground.

Lastly we may cite the *naven* ceremonies as a further example of this proneness to visual and kinaesthetic thought. We have seen how the abstract geometrical properties of the kinship system are here symbolised in costume and gesture; and we may note this in passing as a contribution of eidology to our understanding of the ceremonies.

But the connection between the expression of eidos in the contexts which I have described, and the culture as a whole, is still not perfectly clear. I have illustrated the eidos chiefly from the totemic debating, and have shown that very great activity of memory is demanded and promoted in certain individuals by the sport of debating. Further, I have given facts which indicate that rote memory plays only a small part in this activity, while visual and kinaesthetic imagery appears to be important. In the special business of memorising names I showed that it is possible to suppose that word association plays a part. But these facts might well be isolated in their effects. On the one hand, the active cultivation of memory might be confined to a few selected specialists and, on the other hand, it might occur only in the special contexts in which names are important. Until these two possibilities have been examined, we cannot step from the facts given to the statement that the active development of memory has affected the culture as a whole and the *naven* ceremonies in particular.

We will first consider how far this activation can be supposed to have affected the whole community, and how far it is confined to a small minority of specialists. On the whole, the remarkable keenness in memorising names is to be found in the majority of the men.[1] When I was collecting the names I got my material as far as possible from specialists, but it was noticeable that, even when I was talking about other matters with informants who would never have dared to pose as erudite in public debate, they would continually bring the talk round to matters connected with the totemic system and would attempt to give me lists of names. This was true, for

_____

[1] About the women I have no data, but from the contrast in ethos between the sexes I should expect to find considerable differences in their eidos.

example, of the informant (Plate XXII) whom I have described above (p. 161) as conspicuously enthusiastic and inaccurate. He insisted upon discussing esoterica and giving lists of names belonging to his clan, full of blunders and contradictions. In the younger men, however, this passion for showing off even weakly developed erudition is almost completely checked by the feeling that erudition is only appropriate in senior men. I had three very intelligent youths who consistently avoided giving me names, and who referred me to their seniors when I pressed them. But I was told by other people in their absence that two of these youths were already well on the way towards erudition, and would be great debaters when they were older. Thus the reticence of the younger men on the subject of names does not imply that they are not, like their seniors, keen on this form of mental virtuosity.[1]

But a more complete answer to the question of how the stimulation of a small number of specialists can react on the culture as a whole is provided by the fact that these specialists constantly set themselves up as unofficial masters of ceremonies, criticising and instructing the men who are carrying out the intricacies of the culture. Their voice is heard not only in the debates which concern totemic names, but also in those on every subject from initiation to land tenure. Thus the culture is to a great extent in the custody of men trained in erudition and dialectic and is continually set forth by them for the instruction of the majority. From this we may be fairly certain that the individuals most affected by the stimulation of memory actually contribute very much more than their fellows to the elaboration and maintenance of the culture.

Lastly, there is the question whether this business of totemic names constitutes a separate context, a special interest,

[1] The reticence of the younger men on the subject of names may be compared with their comparative quietness in the ceremonial house, a matter which I mentioned in the discussion of the men's ethos. I said there that in the junior ceremonial house the boys ape the violence of their seniors, but I do not know of any debating in the junior ceremonial house in which the erudition of the older men is copied. There are, however, a number of games in which children test each other's knowledge, e.g. of species of plants in the bush, etc.

the effects of which might be isolated from the culture as a whole; or whether it is a widely ramifying system which might be expected to have effects in all parts of the culture. Here the indications are very strongly in favour of the latter view. The naming system is indeed a theoretical image of the whole culture and in it every formulated aspect of the culture is reflected. Conversely, we may say that the system has its branches in every aspect of the culture and gives its support to every cultural activity. Every spell, every song—even the little songs which boys make up for their sweethearts—contains lists of names. The utterances of the shamans are couched in terms of names. The shamanic spirits which possess the shamans are themselves important nodal points in the naming system. Marriages are often arranged in order to gain names. Reincarnation and succession are based upon the naming system. Land tenure is based on clan membership and clan membership is vouched for by names. The man who buys names acquires at the same time membership in the clan to which the names belonged, and a right to cultivate the land of that clan. Every product of the river and gardens has its place in the system. The land, the river, the sky and the islands of floating grass are all named in the system. Every utensil of the house, the house itself and the ceremonial house are included. Every man, woman and child is included, and the various aspects of social personality are differentiated in the system (cf. *infra*).

Indeed the only province of the culture which is almost independent of the naming system is that of initiation;[1] and

---

[1] Here there is a difference between the Eastern Iatmul (Mindimbit, etc.) and the Central Iatmul (Kankanamun, etc.). Among the latter, both the initiatory moieties and the totemic moieties are called "Sun" and "Mother", in spite of the fact that several clans, which in totemism belong to one moiety and speak on that side in debates about names, have gone across to the other moiety for initiatory purposes. But in Mindimbit, the initiatory dual division is regarded as entirely distinct from the totemic and the initiatory moieties are not named "Sun" and "Mother". They are called Kishit and Miwot. Moreover in Mindimbit the secret slit gongs are not identified with shamanic spirits. In Kankanamun, both gongs and spirits are called *wagan*; but in Mindimbit while the gongs are still called *wagan*, the spirits are called *lemwail*. Thus it would seem that the connection between totemism and initiation is weaker in Mindimbit than in Kankanamun.

even here we find that all the mysterious sound-producing mechanisms have totemic names, the secret slit gongs are vaguely equated with shamanic spirits, and the initiatory moieties have polysyllabic names on the usual plan, though no place has been allotted to these in the distribution of names among the clans. Lastly, the novices receive at the end of initiation special names from their mothers' brothers.

From this extraordinary ramification of the system we may, I think, be fairly certain that the more learned men approach every cultural activity with the cognitive habits acquired in debate; that not only is their memory activated by the culture in the special context of debating, but that this activation has its effects on the culture as a whole, promoting and maintaining the extreme complexity which we have seen to be characteristic.

Hitherto, in describing Iatmul eidos, we have dealt with its quantitative aspects and with questions as to what types of thought mechanisms are most cultivated. It now remains to consider in more detail the patterning of the standardised thought and the sorts of logic which are characteristic of it. Unfortunately I have very little material which would demonstrate the methods of thought of individual natives, and must therefore depend almost entirely upon the details of the culture, deducing therefrom the patterns of thought of the individuals. Ideally it should be possible to trace the same processes in the utterances of informants and in individual behaviour in experimental conditions as well as in the norms of the culture.

In the first place, we must note that a great deal of Iatmul thought is "intellectual"—and here I am using the term in that slang sense in which it is used to imply that people are "intellectual" rather than intelligent.

The problems which most exercise the Iatmul mind appear to us fundamentally unreal. There is, for example, a standing argument between the Sun moiety and the Mother moiety as to the nature of Night. While the Sun people claim Day as their totemic property, the Mother people claim Night and have developed an elaborate esoteric rigmarole about mountains meeting in the sky, and ducks, and the Milky Way to explain its existence. The Sun people are contemptuous of this and Night has become a bone of contention. The Mother people maintain that Night is a positive phenomenon due to the overlapping mountains, etc., while the Sun people maintain that Night is a mere nullity, a negation of Day, due to the absence of *their* totem, the Sun.

Again, within the Sun moiety itself there is matter for dispute about the Sun. One of his totemic names is Twat-mali,[1] but there are two clans in this moiety who lay claim to this name. Each clan has its own string of names linked with this. Thus clan *A* claims that the series runs: Twat-mali, Awai-mali; Ka-ruat-mali, Kisa-ruat-mali; etc. (nine pairs of names), while clan *B* claims that the series runs: Twat-mali, Awai-mali; Ndo-mbwangga-ndo, Kambwak-mbwangga-ndo; etc. (eight pairs of names). At some time in the past a settlement of this argument has apparently been reached in a compromise: that there are two Twat-malis, one of whom is the sun who shines nowadays, while the other is the old sun, who lies as a decaying rock somewhere in the plains north of the Sepik River. But the settlement is only partial, because it has never been agreed which of the two Twat-malis is which, and nowadays each clan taunts the other by saying that their own Twat-mali is in the sky, while that of the opposition lies rotting in the plains.

Another subject which is matter for this characteristic intellectual enquiry is the nature of ripples and waves on the surface of water. It is said secretly that men, pigs, trees, grass—all the objects in the world—are only patterns of waves. Indeed there seems to be some agreement about this, although it perhaps conflicts with the theory of reincarnation, according to which the ghost of the dead is blown as mist by the East Wind up the river and into the womb of the deceased's son's wife. Be that as it may—there is still the question of how ripples and waves are caused. The clan which claims the East Wind as a totem is clear enough about this: the Wind with her mosquito fan causes the waves. But

---

[1] Other clans of the Sun moiety have other names for the Sun—Ianggun-mali, Kala-ndimi, etc. The dispute here discussed is concerned merely with Twat-mali.

other clans have personified the waves and say that they are a person (Kontum-mali) independent of the wind. Other clans, again, have other theories. On one occasion I took some Iatmul natives down to the coast and found one of them sitting by himself gazing with rapt attention at the sea. It was a windless day, but a slow swell was breaking on the beach. Among the totemic ancestors of his clan he counted a personified slit gong who had floated down the river to the sea and who was believed to cause the waves. He was gazing at the waves which were heaving and breaking when no wind was blowing, demonstrating the truth of his clan myth.

On another occasion I invited one of my informants to witness the development of photographic plates. I first de-sensitised the plates and then developed them in an open dish in moderate light, so that my informant was able to see the gradual appearance of the images. He was much interested, and some days later made me promise never to show this process to members of other clans. Kontum-mali was one of his ancestors, and he saw in the process of photo-graphic development the actual embodiment of ripples into images, and regarded this as a demonstration of the truth of the clan's secret.

This intellectual attitude towards the great natural phe-nomena crops up continually in conversations between the anthropologist and his informants, the latter striving to pump the anthropologist about the nature of the universe. I learned very soon that the correct attitude to adopt in such conversa-tions was one of extreme discretion. I parted with informa-tion only after insisting upon secrecy and then shared the secret as a bond between my informant and myself. With such preliminaries, what I said was treated seriously, but without them it was generally assumed to be exoteric lies. One man came quietly, boasting to me that he knew the European secrets about day and night; that a white man had told him that by day the Sun travelled over the earth and then during the night he returned to the East, travelling back over the sky-world, so that the people in the sky have day while we on earth are having night and *vice versa*—a pretty re-phrasing of the Antipodes in Iatmul terms.

A lunar eclipse occurred while I was in Kankanamun and I discussed this with my informants. I was rather surprised to find them but little interested in the phenomenon. They ascribed it to magic performed by the Tshuosh people. One informant, Iowimet, seemed rather doubtful about this theory, and asked me what we Europeans believed. I hedged for a while, and then explained the secret. The moon during the eclipse had turned red, and my informant had told me some days previously that the sun was a cannibal. I ascribed the redness of the moon to blood in the sun's excrement, and said that this material was between the moon and the earth obscuring the moon's brightness. This secret was quite a useful bond of understanding between my informant and me, and several times later he referred to it as "that matter we both know about". It might be interesting to collect in the future Iowimet's re-phrasing of what I told him.

More specifically, Iatmul thought is characterised not only by its intellectuality, but also by a tendency to insist that what is symbolically, sociologically, or emotionally true, is also cognitively true. Apparently the sort of paradox which can be constructed in this way is very attractive to the Iatmul mind, and the same mental twist is, of course, recognisable in dialecticians and theologians in all parts of the world. Among the Iatmul the dialecticians and theologians are not a class apart but are, as we have seen, the chief contributors to the culture. Thus it comes about that many of the complications of the culture can be described as *tours-de-force* played upon this type of paradox, devices which stress the contradiction between emotional and cognitive reality or between different aspects of emotional truth.

As examples of this, we have the oft-repeated idea of the discrimination between and yet the identity of different aspects of social personality. We have already noted this cultural phenomenon in the naming of human beings; we have seen how a man's various names constitute a diagram of his various allegiances to his paternal clan, his maternal clan, and his classificatory *waus*, and how another name is connected with his status as an initiate. The various souls of a man with their different fates after death—his patrilineal aspect which is reincarnated in his son's son, his matrilineal aspect which dwells on in the land of the dead under the name given him by his *wau*, and that aspect of his personality which is represented by his secret name, which becomes a potsherd spirit, a guardian of his clan—correspond in the same way with the various facets of his personality.

The discrimination of facets of personality is carried even farther in the case of *wagan*, the most important type of spiritual being. Kava-mbuangga, the *wagan* who set his foot upon the mud and thereby created dry land, is a representative both of the living and of the dead, and there are myths of his visit to the land of the dead. A figure of him which I was shown in Mindimbit was painted in two colours, the right-hand side of the figure in ochre to represent living flesh, and the left-hand side in black to represent stone, the *kuva* (paralysis, or "pins-and-needles") which creeps over the body in death. Thus in the figure the two sides of Kava-mbuangga's personality are diagrammatically shown.

In their naming, too, the *wagan* are differentiated, not only *qua* members of their maternal and paternal clans and as initiates, but also they have separate strings of names according as they function as (*a*) shamanic spirits; (*b*) avenging spirits, invoked in sorcery; (*c*) secret slit gongs in initiatory ceremonies; (*d*) *mbwatnggowi* (ceremonial dolls which are vaguely believed to be associated with fertility); (*e*) *mwai* (long-nosed masks in which the young men dance. These dances are junior analogues of the *wagan* slit-gong ceremonies performed by the senior men).

These different names of the *wagan* are many of them esoteric, and especially to be concealed is the fact of fundamental identity, which exists in spite of the differentiation between the various personifications. Only the more erudite men know that *mwai* and *mbwatnggowi* are really *wagan*; and when they told me of this they spoke of it as of a mystery which was beyond their comprehension, a paradox which had to be accepted and at which they marvelled with a certain serious humility and acceptance of its incomprehensibility. At the same time, they insisted upon the identity with an over-emphasis in favour of that belief which, of the two, was the more difficult to assimilate.

Another example of insistence upon a sociological truth because it looks like cognitive nonsense may be seen in the constantly repeated statements that the Borassus palm, Tepmeaman,[1] is a fish as well as a palm. Tepmeaman is the *kaishi* (counterpart or partner) of *na'u*, the sago palm. But, in Iatmul economics, the counterpart of sago is fish, since the river people catch fish which they exchange for sago manufactured in the bush villages. Thus we can detect some justification for the statement that the Borassus palm is a fish.

The tendency to emphasise the more cognitively obscure of two truths is recognisable again in the kinship system. In my preliminary account of the culture I stated that while the "morphology of the social system"[2] is patrilineal, the "sentiment" of the people is preponderantly matrilineal. I am inclined now to see in this almost sentimental stressing of the importance of the tie with the mother another instance of over-compensation in favour of the less evident truth. My informants, when we were discussing the mother and mother's brother, had certainly a trace of that same puzzled insistence which was present when they told me of the different personalities of *wagan*. It would seem then that the patrilineal relationship is evident enough from clan organisation to need no stressing. But its clear existence casts a slight obscurity upon the matrilineal relationship which, though no less socially and emotionally real, is thereby rendered less cognitively evident. Therefore the matrilineal relationship is the more emphasised.

---

[1] The names of this palm are Tepmeaman, Kambuguli, etc. (cf. p. 44, where the complete list is given). In a previous publication (*Oceania*, 1932) I have stated that Kambuguli is a different species of palm from Tepmeaman. On my second expedition I found that borassus is a dioecious palm and that Kambuguli is the male plant as distinct from Tepmeaman, the female. Actually the natives regard Tepmeaman as the male, the word *tshik* being used homonymously for "fruit" and "penis".

[2] *Oceania*, 1932, p. 289. By "the morphology of the system" I meant those cultural premises which are articulately formulated by the natives, e.g. groupings such as clans, moieties, initiatory grades, etc., which are named and discriminated. A man's ties with his father are articulately stated, while those with his mother are chiefly expressed in symbolic behaviour.

Another example of the same sort of thinking occurs in the identification of the *laua* with the ancestors of the maternal clan (cf. p. 43). Here the link upon which the equation is based would seem to be a purely emotional one. In the two relationships, that between a clan and its *laua*s and that between a clan and its ancestors, the only common factor is the pride which is associated with the achievements of both *laua*s and ancestors.

In the *naven* ceremonies we can see the same sort of twisted thinking acted out in ceremonial. In a social and economic and emotional sense, the *wau* is a mother and wife of his *laua*. But in a strictly cognitive sense he is not anything of the sort. In the carrying out of the ceremonial, it is the emotional truth which is stressed, which, of the two, is the more difficult of cognitive assimilation.

Besides this sense of the contrast between emotional and cognitive truth, Iatmul thought has certain other patterns or motifs in which it is trained by the culture and which pervade the various institutions. At this stage in the book it will hardly surprise the reader to be told that these patterns are to a great extent mutually contradictory, so that the culture as a whole appears as a complex fabric in which the various conflicting eidological motifs are twisted and woven together. Unravelling and isolating these motifs, we can recognise and describe some of the more important:

(*a*) A sense of pluralism: of the multiplicity and differentiation of the objects, people, and spiritual beings in the world.

(*b*) A sense of monism; that everything is fundamentally one or at least is derived from one origin.

(*c*) A sense of *direct* dualism: that everything has a sibling.

(*d*) A sense of *diagonal* dualism—that everything has a symmetrical counterpart.

(*e*) Patterns of thought which govern the seriation of individuals and groups; these patterns being apparently based upon (*c*) and (*d*).

The senses of multiplicity and of monism need only very brief treatment, since neither of them, so far as I know, has had any repercussions on the *naven* ceremonies. On the one hand, we have the type of material which I have already mentioned, the distribution of thousands of personal names to the things and people of the world and to the different facets of their personalities; and on the other hand, we have the insistence that everything is really *one*. On the whole, it would seem that the belief in multiplicity is lay and exoteric, while the details of the basis of monism are an esoteric mystery for the erudite.

This does not mean that the general concept of monism is a secret. When two clans are striving over the ancestral ownership of a name, a man may declaim in open debate: "One father, one mother, one root—that is the sort of people that we (human beings) are", insisting, by this conventional reference to one of the pairs of first parents of the human race, upon the indivisibility of the disputed name.

But the names of these first parents are secret, as also are all the details of muddled mythology whereby monism unites the whole world in historical mysticism.

The same considerations apply to *wagan*. There are many *wagan* and many of them have multiple personalities, and yet all *wagan* are one and were descended from Kava-mbwangga who multiplied himself by means of boluses of chewed betel upon which he trod and which thereby became the other *wagan*; but which are all really Kava-mbwangga. Another instance of this mystical monism is to be seen in the theory that all things in the world are only the patterns of waves.

But over against the sense of the unity of everything, we may set the confusion which has perhaps resulted from the rivalry between clans and the emphasis upon pride. Everybody believes and insists on fundamental monism, but each group of clans has its own mythological theory according to which its own ancestors are given the key positions in the unitary origin of the world. Thus it has come about that there is not one monistic theory but a whole series of conflicting theories, each of which emphasises the fundamental unity of the world.

There are at least two (and probably more) pairs of first parents included in the polysyllabic totemic scheme; and besides these there is a third pair who have short names and are attached to no clan[1]—and so the culture swings between monism and pluralism.

An interesting expression of the fundamental pluralism of the world may be seen in the native theory of separate "roads". This is a matter upon which the natives themselves are vague and opinion varies from one informant to another. I was told that human beings, *wagan*, *kurgwa* (witches), and *windjimbu* (wood spirits), have all of them separate "roads". But some informants were inclined to think that there were only two "roads", that of human beings and that of spirits; others again discriminated three roads, that of *wagan* and *kurgwa*, that of *windjimbu*, and that of human beings. I was told in pidgin English that *wagan* were "behind true", i.e. that though invisible they were yet present in some mysterious way which we might express by a reference to the "fourth dimension" or to another "plane of existence". The word, *iamba* (road), as applied to these spiritual beings seemed, although the roads were described as means of instantaneous transport, to be also an equivalent of "planes of existence".

While the monism and pluralism in Iatmul thought are comparatively easy to understand, the two forms of dualism are somewhat more obscure. The matter is, however, of some importance, and I believe that the discrimination of direct and diagonal dualism may prove useful in the description of cultures other than the Iatmul.[2]

---

[1] It is interesting that my informant, Kainggenwan of Mindimbit, who gave me the names of these two non-totemic first parents, Mogavia (the man) and Leren (the woman), regarded these names as secrets of great importance; although as they have no clan status and no magical efficacy we might expect the names to have no value.

[2] This exposition of direct and diagonal dualism was written before I realised that these concepts are only eidological analogues of complementary and symmetrical ethos. The phrasings which are here used are in some respects clumsy, and some simplification might have been introduced by substituting the term *complementary* for "direct" and *symmetrical* for "diagonal". It seemed worth while, however, to leave the original statement of the matter almost unchanged so that the reader might see how schismogenic phenomena were reflected in my account of Iatmul eidos *before* I had developed the concept of schismogenesis. The equivalence between the two types of dualism and the two types of schismogenesis is worked out more fully at the end of the book, since it was only in the writing of the Epilogue that this identity became clear.

By dualism I mean a tendency to see things, persons, or social groups as related together in pairs; and according to the type of relationship which is seen between the elements in each pair, I shall speak of the dualism as being either *direct* or *diagonal*. In the case of direct dualism the relationship is seen as analogous to that which obtains between a pair of siblings of the same sex; while in diagonal dualism the relationship is seen as analogous to that between a pair of men who have married each other's sisters.

These considerations were suggested to me by a paper read by Dr Leonard Adam,[1] who mentioned a Nepalese institution which he described as "artificial brotherhood". The existence of two forms of dualist thinking became evident from considering the contrast between this relationship and the corresponding "artificial" relationships institutionalised in Iatmul culture.

In almost all cultures there are to be found these institutionalised partnerships, and in the simpler societies it is usual to find that these partnerships, contracted either outside or across the bonds of kinship, are yet patterned upon the familiar kinship ties. Thus in Nepal the behaviour of the partners is apparently patterned upon that normal between brothers in Nepalese culture. Among the Iatmul, on the other hand, there is no "artificial brotherhood", but in its place we find a series of types of "artificial" relationships patterned upon affinal ties and stressing the symmetrical nature of these ties.

The contrast is very striking and led naturally to the problem of what cultural factors could be expressed by institutionalising the one type of "artificial" relationship rather than the other. The theory which I am now putting forward is an attempt to answer this question from an eidological point of view.

The theory is simply that there are two contrasting types of dualist thinking. The first type, which is easy for Euro-

---

[1] *Man*, vol. xxv, 1935, p. 12 and *Zeitschr. für Vergl. Rechtswissenschaft*, Bd. xlix. For instances of "artificial relationships" probably based upon diagonal dualism in African communities, cf. Hocart, "Blood Brotherhood", *Man*, 1935, p. 127.

peans to follow, would classify together the adjacent corners of a rhombus (to employ a geometrical analogy). This type of thought, which I shall refer to as, *direct* dualism, is that which leads to the formation of artificial relationships based upon brotherhood and to concepts such as that of the Iatmul that everything in the world can be grouped in pairs, such that in each pair one component is an elder sibling, while the other is a younger sibling of the same sex.

The second type of thought is much more difficult for Europeans because it is less developed in our eidos.[1] It would classify together the diagonally opposite corners of a rhombus, stressing their fundamental equality and oppositeness. This diagonal way of thinking leads in Iatmul culture to the formation of artificial affinal relationships and to the idea that everything in the world has its equal and opposite counterpart. It is extended farther to the great dualist constructs underlying the two moieties of Sun and Mother with their opposed totems Sky and Earth, Day and Night, and to the cross-cutting dual divisions of the initiatory system in which one half of one moiety initiates the diagonally opposite half of the opposite moiety (see Fig. 4, p. 245).

It is possible that these two patterns of thought are both of them represented in the eidos of all cultures, but I believe that there is a great deal of variation between cultures in the degree to which one or the other type of dualism is stressed. Among the Iatmul, both types are strongly developed. It is hard to estimate their relative importance because the diagonal way of thinking appears more strange and striking to the European observer. My impression is that greater emphasis is given to the diagonal type, and further that this emphasis is stronger among the Eastern Iatmul (Mindimbit, Tambunum, etc.) than among the Central Iatmul (Palimbai, Malingai, etc.). In Mindimbit I collected two sorts of lists of names of shamanic spirits arranged in pairs. In one list

---

[1] It may reasonably be asked why, since both kinds of schismogenesis are present in European society, it should be more difficult for us to think in terms of diagonal (i.e. symmetrical) dualism. The answer is that while we habitually think of complementary relationships in dual terms, we do not think of rivalry and competition as necessarily dual. Our society, though it is based upon competition and rivalry, lacks bilateral symmetry.

the pairs were composed of elder brother and younger brother names, while in the other list they were pairs of *kaishi* (opposite counterparts). In Palimbai I tried to collect the same two sorts of lists and was only given lists arranged in sibling pairs. When I asked for the names paired as *kaishi* and said I had been given such lists in Mindimbit, my informant said, "Yes, the Easterners are always talking about *kaishi*; with them, everything has a *kaishi*; but we are not like that." This eidological difference between Palimbai and Mindimbit is in part due no doubt to historical accident, and it is interesting that the Easterners are geographically nearer to the area in which extremely complicated symmetrical kinship systems have been collected among the Banaro[1] and Mundugumor.[2]

But in spite of this difference in degree between Palimbai and Mindimbit, both types of dualist thought are well developed throughout the Iatmul area; and both are recognisable in the shaping of *naven*. On the one hand we have the emphasis upon the various direct identifications of siblings, of father and son, and of husband and wife, and on the other hand, as examples of the diagonal way of thinking, we have the whole patterning of the brother-in-law relationship and the beautiful reversed symmetry in *naven* on the two sides of the marriage tie, whereby the father's sister dresses up as a man and identifies with the father, while the mother's brother dresses up as a woman and identifies with the mother.

There are, of course, other types of dualism besides the two which have been cultivated by the Iatmul. In our own culture, we have the dualism between Mind and Matter and others based on contrast between different levels of the personality. In these types of dualism the Iatmul have, so far as I know, not indulged; neither have they emphasised at all strongly any dualism based upon contrasting ethological patterns. This is perhaps surprising in a culture in which each sex has its own ethos. Even the common idea of "male" and "female" principles in nature is only recognisable as a very minor motif in the totemic dual division. The Sun is

[1] Thurnwald, *Mem. Amer. Anth. Ass.* vol. III, 1916.
[2] Margaret Mead, *Sex and Temperament*.

personified as masculine and the Moon, which belongs to the
Mother moiety, is a woman. The Mother moiety has the
vulva (not, so far as I know, personified but feminine in
gender) as its most important emblem. But this type of
contrast is not carried far, and the earth and sky are both
personified as masculine. In general, Iatmul thought has
ignored the sort of philosophy which finds ethological dif-
ferentiations in nature. There has never been a Blake among
them; and their myths of origin of the world emphasise the
separation of sky and earth rather than the coming together
of great opposites to produce the world. Indeed there is so
little tendency to link ethological contrasts on to diagonal
dualism that with some hesitation I have classified the
identification of man and wife among those based upon direct
dualism instead of putting it in the same class with the
identification of brothers-in-law. In marriage, the contrast
in sex ethos makes impossible the element of symmetrical competition which characterises co-operation between brothers-
in-law.[1]

Lastly, before leaving the questions of direct and diagonal
thought, we may note that primitive peoples, as well as
anthropologists, tend to confuse the two methods of thinking.
In a typical classificatory system, the distribution of kinship
terms is based upon the identity between siblings of the same
sex. Thus the use of identical kinship terms for two different
persons becomes typically a device for stressing their direct
social identity, but the same device may also be used to stress
identity based upon diagonal thinking. Thus among the Sulka

---

[1] This paragraph was written *before* I had equated direct dualism with
complementary patterns, and I have left the passage unchanged. I had
found that those with whom I discussed the types of dualism tended to
confuse the geometrical oppositeness of diagonal dualism with the sort
of oppositeness which occurs in the contrast between Day and Night.
My concepts were at the time rather vague but I was sure that this was
a misconception, and this passage is therefore devoted to showing that
oppositeness of nature is not implied in diagonal dualism. It did not
occur to me to put the matter as I now should in positive terms, stating
that complementary nature *is* implied in *direct* dualism.

The lack of emphasis in the totemic system upon "male" and "female"
principles is probably correlated with the fact that this system is a reflection of the subdivision of the community into moieties and clans rather
than of the differentiation of the sexes.

of New Britain, whom I studied for some months in 1929, the kinship term used for own mother's brother is extended to own mother's brother's counterpart (*krus*) in the opposite moiety, a man whom the speaker would address as "father" were it not for his diagonal identification with the mother's brother. In the Iatmul system the same phenomenon is recognisable in the observance by father's *tambinyen* of the taboo on eating food collected by the son, a taboo which chiefly concerns the father.

Again, a confusion of direct and diagonal dualism occurs in the system of totemic moieties and clans. This system is thought of as a series of diminishing dichotomies, but it is not perfectly clear whether these dichotomies are direct or diagonal. I have already indicated that the relationship between the moieties is, on the whole, diagonal. But pairs of related clans are thought of as descended from pairs of elder and younger brothers and thus the relationship between them is conceived of as direct; and this concept has been extended to the moiety, so that it is said by the Sun people that their moiety is "elder brother" of the Mother moiety.[1]

Equally, instead of stating that the direct dualism which occurs between pairs of clans has been extended to the relationship between the moieties, we may say that the diagonal dualism between moieties has influenced the relationship between the clans. The lists of totemic ancestors of the various clans show that a great deal of imitation has occurred. Every clan has its personified pig, eagle, fire, crocodile, ceremonial house, shamanic spirit, etc.; and it is likely that the sense of symmetry connected with diagonal dualism has guided the various clans in their copying of each other (i.e.

---

[1] I think I was also told by members of the Mother moiety that their group was "elder brother" of the Sun moiety but I have no record of this in my notes. I probably took but little notice of the statement when it was made, ignoring it as a boast. But the matter is interesting, since if difference in age, the mark of complementary relationship, is become a matter for symmetrical rivalry, this might throw some light upon the alternating system of grouping of initiatory grades into moieties with symmetrical rivalry between them. Cf. also the quarrel between Kwoshimba and Kili-mali (p. 105) who were related symmetrically through the marriage of their children. Kwoshimba finally boasts that he is Kili-mali's "elder brother".

that the relationship between clans though nominally complementary and phrased in terms of "elder" and "younger"
is to some extent patterned upon symmetrical rivalry).

There is one peculiarity of the Iatmul handling of direct
dualism which needs to be specially emphasised (and from
this, more than from any other detail I shall in the Epilogue
deduce an equivalence between "direct" dualism and complementary patterns of relationship). Every thing and every
person has a sibling and the polysyllabic names are so arranged in pairs that in each pair one name is the *elder* sibling
of the other. Throughout the whole field in which direct
dualist thought is recognisable, it is accompanied by the concept that one of the units is senior to the other. There is no
such concept in the case of persons identified by diagonal
dualism and it would seem that such persons are nominally
equal in status and always of the same sex; while those who
are directly identified can never be equal but must differ
either in seniority or in sex.

In the case of siblings of opposite sex, there is no emphasis
on seniority, the difference in sex being apparently analogous
to difference in age. A sister, regardless of age, always
addresses her brother as *nyamun*, the term used by a man for
his elder brother only. But between siblings of the same sex,
the terms *nyamun* (elder sibling of same sex) and *tshuambo*
(younger sibling of same sex) are always employed. In the
naming system, there are pairs of masculine names and in
many cases corresponding pairs of feminine names, thus:

Mwaim-nanggura-ndimi	Mwaim-nanggura-ndimi-ndjowa
(elder brother)	(elder sister)
Temwa-nanggura-ndimi	Temwa-nanggura-ndimi-ndjowa
(younger brother)	(younger sister)
⋮	⋮

etc. (nine pairs of names). etc. (seven pairs of names).

Always the seniority is stated in the same sex pairs and
omitted between the sexes.

In other cultures in which direct dualist thought has been
at work, we may find it expressed in mythological references
to twins. But this type of myth is, so far as I know, absent

from Iatmul culture. It is likely that the extreme horror and intolerance of twins in some cultures are a further development comparable with the Iatmul stressing of seniority. Among the Iatmul, twins are tolerated and no great interest is taken in the phenomenon of twinning. There are, however, a few myths which embroider upon the idea of namesakes (*oiseli*). One man meets another and on asking his name finds it to be the same as his own. He then asks the father's name and finds (as the naming system theoretically demands) that it is the same as that of his own father. In mythology namesakes immediately strike up a friendship. But in real life they seem to pay but little attention to each other.

Leaving now the instances of paired identification or opposition and turning to longer series, a new pattern of thought is recognisable. In Iatmul culture, with its lack of chieftainship or hierarchy, there are but three contexts in which such seriation occurs,[1] and in all of them the first, third, fifth, etc. units are grouped together, and this group is contrasted with that consisting of the second, fourth, sixth, etc. The three contexts in which this patterning is developed are: the alternation of generations, the alternation of initiatory grades, and the alternation of siblings.

A man's own generation, his paternal grandfather's generation, and his patrilineal grandson's generation, are grouped together as one *mbapma* (literally "line"), and in contrast to this his father's and his son's generations are of the opposite *mbapma*. I have already mentioned that theories of reincarnation are linked up with this alternation and I need only add here that a man may address his father's father either as *nggwail* (grandfather) or as *nyamun* (elder brother). This identification of relatives with others in analogous positions two generations away on a patrilineal line runs through the whole kinship system, so that a man's son's wife is *nyame* (mother) and his son's wife's brother is *wau* (mother's brother).

The initiatory system, the second context in which the

---

[1] There are one or two minor instances of seriation without grading, e.g. the planting of yams in rows and the arrangement of house posts. In these instances there is no alternation.

same sort of alternation occurs, can only be described very briefly here.

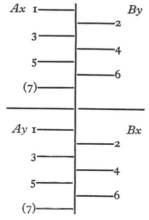

Figure 4. Diagram of Initiatory Groups.

The two cross-cutting moieties are marked *A* and *B*, and *x* and *y*. Thus the resulting quadrants are *Ax* and *Ay*; *By* and *Bx*. Members of a moiety initiate youths of the opposite moiety. Thus the members of *Ax* initiate *By*, and *Ay* initiate *Bx*. Each quadrant is divided into three named generation groups: 1, 3, 5 or 2, 4, 6; such that 1 are the fathers of 3, who are the fathers of 5; while 2 are the fathers of 4, who are the fathers of 6.

The system is so arranged that one of the moieties, *A* or *B*, is always senior to the other. In the Diagram, *A* is represented as senior to *B*. Thus, of the generation groups in *A*, each is slightly senior to the corresponding group in *B*. Members of *Ax* 1 initiated the members of *By* 2. *Ax* 3 are now initiating *By* 4. *Ax* 5 are only just ready for initiation but will later be initiated by *By* 4. *By* 6 are all small children. Similarly, *Ay* 1 initiated *Bx* 2, etc.

In course of time, *Ax* 1 will die off and a new generation *Ax* 7 will appear. The moiety *B* will then become senior to *A* and their seniority will be recognised when they assert their strength violently in a brawl. *By* 4 will then begin initiating *Ax* 5.

In this beautiful system the groups 1, 3, 5 are united, and opposed to 2, 4, 6, which alternate with them and are always their junior by about half a generation. Thus the pattern of thought which appeared as an alternation of generations in the patrilineal clan here appears as an alternation of half-generation groups which are staggered like the spokes of a wheel.

Throughout the system each grade is spoken of as the "elder brother" of the grade next below. Thus 1 is elder brother of 2, and 2 of 3, etc. This usage links up with the third context in which the same patterning occurs.

In large families in which there is a long series of brothers, the same sort of alternation is expected. The language possesses five terms (not numerals) for first, second, third, fourth and fifth child respectively; and it is expected that, in their quarrels over the patrimony in which they share before the father dies, the first and third brothers will join forces against the second and fourth. I asked about the fifth brother but was told "he would be too little, he would watch them".

Lastly there is one other context in which I am tempted to see a development of the same type of pattern, namely in the structure of flute music.[1] The initiatory flutes have no lateral stops by which the player might control and vary the pitch. He is therefore limited to the series of natural harmonics provided by the tube. A flute will give about seven notes distributed over three octaves, but the intervals between the notes are too large to admit of any tune being produced by a single flute. This difficulty is overcome by a scheme of alternation. The flutes are always played in pairs and in each pair one flute is one tone higher than the other. In the playing, as a result of the difference in pitch between the two instruments, the harmonics of the one alternate with the harmonics of the other; so that the two performers, who blow in turn, can together produce simple tunes. The flute of lower pitch is spoken of as the "elder brother" of the other, though for totemic purposes the pair of flutes is taken as a unit.

Whether this staggering of the harmonics of flutes is really another expression of the sort of thinking with which we are here concerned may perhaps be doubted. But from the alternation of generations, initiatory grades, and brothers,

[1] *The Eagle*, St John's College Magazine, 1935, p. 258.

the reality of the phenomenon is clear enough. So far as I know, this pattern of thought is not recognisable in the shaping of *naven*, apart from such small points as the identification of man's sister's husband's father as a *laua* (cf. distribution of pigs, p. 18). But I have described this alternation in some detail because of its general interest and because I believe that we may see in it a combination of the two types of dualist thinking, an intrusion of diagonal thought into contexts where direct identification and the principle of seniority were already at work. Thus between pairs of brothers only these two factors are concerned, but in longer series we find the principle of symmetrical oppositeness breaking up the series.

In this connection it is interesting to compare the myth of the Dioscuri, Castor and Pollux, with the alternations characteristic of Iatmul eidos. Of these twins, Castor was mortal and Pollux was immortal. The mortal twin was finally killed in battle and his immortal brother then interceded with Zeus who granted him, as a favour, that he should share his immortality with his brother. From that time, they have existed on *alternate* days.[1]

The resemblance between the alternation of the Heavenly Twins and that of brothers and initiatory grades among the Iatmul is striking; but we may cautiously ask what, if anything, it signifies. In the first place it is obvious that an eidological resemblance between two cultures need have no significance from a sociological or ethological point of view. Still less will it imply any historical connection between the cultures. The resemblance between the two systems has only been demonstrated within strictly defined limits and we have no right, when we draw our conclusions, to wander outside these limits into realms of history, sociology, or economics. We cannot rush on to deduce from the eidological resemblance that the relationship between brothers among the Greeks and the Iatmul is alike in any single pattern of

---

[1] Other versions of this story occur and are less comparable with the alternations of Iatmul culture. It is said, for example, that Pollux was permitted to spend his days alternately among the gods and with his brother in Hades.

cultural behaviour. It is obvious that the same eidos might be expressed in the two cultures by means of very different forms of behaviour.

The only conclusion then which we are entitled to draw is that comparable patterns of thought occur in the Greek and Iatmul cultures, and we may note that in both cultures these patterns are expressed in contexts of the relationship between brothers. Further than this we cannot go. But if the strict isolation of eidological, ethological and sociological points of view permits us even this much freedom in comparing the details of one culture with those of another I believe that it will be worth while. Hitherto we have had no criteria of equivalence which would indicate either what comparisons might be permissible or what sort of conclusions we might base upon the resemblances which we found.

I suggest that, if we discipline ourselves to think rigidly in terms of one sort of social or cultural function at a time, we shall be free at last to compare a detail from one culture with a detail from another, knowing what we are comparing and what sort of deductions may be made from the comparison. It is true that our conclusions will be so limited in scope as to appear almost useless, but at least they will have one use, that of suggesting further problems. While eidological conclusions can throw no light upon other aspects of cultural behaviour, they may at least suggest problems, the answers to which must be sought in terms of these other aspects. The detection of similar eidos in two different cultures may well stimulate an enquiry into the sociology or economics of the peoples concerned.

Another comparative point which will almost certainly repay investigation may be seen in the resemblance between the Iatmul kinship system and the class systems of Australia, Ambrym, etc. In all these systems we may recognise both diagonal and direct dualist thought, and in all of them we find an emphasis upon the alternation of generations. One conspicuous difference, however, between the Australian systems and that of the Iatmul is that the former are *closed*. An Australian community is divided into a fixed number of groups, and it is laid down which group shall marry which group and which generation shall marry which generation. This is not so in the Iatmul system, and even if the Iatmul confined themselves more rigidly to *iai* marriage they would not have a closed system like those of Australia. Here, I believe that we are concerned with a rather important motif of Iatmul thought. The natives see their community, not as a closed system, but as an infinitely proliferating and ramifying stock. A clan will grow big and it will subdivide; a village will grow big and it will send out colonies. The idea that a community is closed is probably incompatible with this idea of it as something which continually divides and sends out offspring "like the rhizome of a lotus".

It is worth speculating about another difference between Australian kinship and that of the Iatmul. It would appear from the accounts of Australian kinship that these people think of the assignment of kinship terms as determined by the system of social groups; in fact, that in Australia classificatory kinship is really *classificatory*. In contrast to this, the Iatmul certainly assign kinship terms rather on a basis of *extension* from the family itself. We are not concerned here with the question of how in the dim past such and such a term came to be assigned to such and such a relative, but rather with the purely synchronic question of what method of thought the people use now in assigning kinship terms to their various relatives, and it is clear that in a classificatory system it is possible to argue either from the social groups to the family or from the family to the social groups. It is probable at least that some peoples do one while other peoples do the other. In the case of the Iatmul, I should say that this people assigns terms upon a basis of extension rather than of classification.

This is a matter about which there has been a certain amount of discussion in anthropological circles, and it would appear that some anthropologists are prone to one way of thinking, while others are prone to the other. It would be important to know whether some native peoples differ from others in this same respect. It is likely that some cultures promote the inductive methods of thinking, while others promote the deductive.

If there be a possibility that the ways of thought of anthropologists occur, too, among primitive peoples, what then are we to say of the various ways of thinking which have been advocated in this book? We may profitably at this stage go back over the system which we have constructed and consider what it means in terms of native eidos. We have documented the fact that every piece of behaviour is relevant (*a*) to society as a whole, (*b*) to the emotions of the individual, and (*c*) to the thought of the individual, and we have trained ourselves to think separately in terms of each one of these forms of relevance. In addition to these forms of relevance there are others which we might have studied: notably we might have considered the whole culture as a mechanism oriented towards the production, distribution and consumption of material objects, or again as a mechanism which moulds the personalities of the individuals.

The suggestion which I now wish to put forward is that native peoples themselves are more or less conscious of these various aspects of their behaviour, and that cultures may differ profoundly in the extent to which one or the other aspect is emphasised[1] in the consciousness of the individuals.

If we ask an individual why he behaves in a particular way, there are certainly five types of answer which he may give. He may say:

1. "This behaviour promotes the well-being of the community", answering our question in sociological terms.

2. "This behaviour is customary, or traditional, or consistent with other behaviour", answering our question in structural terms.

3. "This behaviour gives me certain types of emotional satisfaction"; or "it gives emotional satisfaction or dissatisfaction to other people, and I enjoy that", answering our question in ethological terms.

---

[1] It is doubtful, perhaps, to what extent this sort of standardisation should be regarded as either ethos or eidos, but I do not think that this question need bother us materially at present. It is possible that some other term should be introduced for these various sorts of standardisation, but the syndromes associated with them have not yet been sufficiently investigated to justify the introduction of further terminology.

4. "This behaviour will supply such and such useful objects", answering our question in economic terms.

5. "This behaviour will train my mind, or somebody else's mind", or "it results from my upbringing", answering our question in developmental terms.

But, as we have seen, at any rate in the case of the first three points of view, all the sorts of relevance are actually present in all behaviour.

The question which we have to decide is, granted that all the aspects are present, Is the individual equally conscious of them all? I would suggest that in different cultures these different aspects are stressed in varying degrees, and that a profound difference appears in a personality according as one or the other aspect of behaviour is chiefly conscious.

In enquiring into this matter, however, one precaution is very necessary. The reader will have noticed that a certain effort of mental acrobatics is required to shift the mind from thinking in terms of one point of view to thinking in terms of another, and further that he himself thinks more readily in terms of one or another of the points of view. I myself tend, at least in conscious levels of my mind, to think most readily in structural terms; and in estimating the relative emphasis in Iatmul culture of these different methods of thought I may have been influenced by this bias. It is probable that the questions which I asked of my informants were so phrased as to indicate that I expected an answer in structural terms.

We have seen that in Iatmul culture, sociological concepts are usually expressed in symbolic terms. The well-being and fertility of the community is ascribed to the spiritual beings called *mbwan* or *kop*; the fighting force of the village is symbolised by the eagle; and the unity of the initiating grade is represented by the crocodile.

A more interesting series of sociological and other phrasings is provided by native statements about the vengeance sanction (cf. p. 55). We have seen that there are various phrasings of the matter. We have first the phrasing in terms of *ngglambi*, the dark cloud which is seen over the house of the guilty man and which may cause the sickness of his relatives. This phrasing with its loose inclusion of the whole group of the sinner's relatives we may regard as sociological. Secondly, we have the phrasing in terms of *wagan*, and this too would seem to be a sociological phrasing, differing from the first in that the group of individuals whose unity is here symbolically represented is, not the group of the relatives of the offender, but the clan of the offended man. It is the *wagan* of that clan who avenges the offence against one of the clan members. Thirdly, we have the phrasing in terms of individual vengeance or *lex talionis*. Here the matter is stated simply in terms of the emotions and motives of individuals and the symmetry of their behaviour; while the symbolism characteristic of the more sociological phrasings has disappeared.

In all such sociological phrasings we may recognise the development of a system of thought which fits in well enough with the theory of religion put forward by Durkheim, and we may suspect that inasmuch as sociological phrasings tend to be symbolic, this aspect of behaviour is not one which is emphasised in Iatmul consciousness.

It is interesting in this connection to consider how in modern European "totalitarian" communities, whether they be Fascist or Communist, an aspect of behaviour which was formerly almost ignored is now stressed by propaganda of every sort. The individual is made to see his behaviour more and more as relevant to the State. Conscious phrasing of the sociological motive in terms of the State seems to replace a former symbolic phrasing of the matter in terms of a deity.

Structural phrasings, on the other hand, seem to me to be much more developed in Iatmul culture. The *wau* is articulately stated to be a mother, and much of his behaviour is seen by the natives as consistent with this major premise. A great deal of cultural behaviour, too, is seen by the natives as "traditional", and this, I think, is probably an expression of a structural point of view. The statement is one which involves no elaborate syllogism: it amounts on examination merely to saying that a thing is done because it has been done in the past. But this is, after all, a statement of structural consistency.

Another feature of Iatmul culture which is probably an expression of structural thinking is the native tendency to pithy clichés. "She was a fine woman"—so they married her inside the clan. I do not know exactly to what extent phrases of this sort are clichés or proverbs, but they were enunciated by my informants with an intonation quite different from that which would accompany a statement of the motives for behaviour.

Another example of the Iatmul interest in the structural aspects of behaviour may be seen in their insistence upon the schematic qualities and the straightness of their marriage system (p. 91) and of their lists of totemic names (pp. 127, 128).

We may suspect that this emphasis upon the structural viewpoint is linked up with a lack of emotional ease which we have seen to be characteristic of the Iatmul men and, if this correlation be correct, then there is a possibility that a structural view of the world is an important symptom in the syndrome which Kretschmer has called schizothymia; and questions arise as to the possibility of schismogenesis between personalities which emphasise one of these points of view and personalities which emphasise another.

Phrasings in terms of emotion are, on the other hand, rather rarer among Iatmul men, and the only two which I can recall are phrasings of unpleasant emotion—*ngglangga* and *kianta*—the former a reference to the emotion of a man whose pride has been hurt, which might perhaps be translated as "pique", the latter a term for jealousy. Both of these terms are occasionally used in stating the reasons for behaviour, but I never heard any reference to pleasant emotions as the cause of any detail of behaviour. The term *wowia kugwa*, "to be in love", is used almost solely of women, and it is likely that among the women emotional phrasings of reasons for behaviour are very much more frequent than among the men.

Economic phrasings occur not infrequently in the culture. But though records are kept of every bride price, and though a Iatmul man is very capable of driving a hard bargain, there is in this Papuan culture none of the enormous economic emphasis which is so characteristic of the majority of Mela-

nesian peoples. Currency and trade and the accumulation
of wealth could never be described as the major pre-occupa-
tions of the Iatmul. It might be worth enquiring into the
relationship between cultural emphasis upon economic aspects
of behaviour and the type of personality called anal in psycho-
analytic jargon.

Another point of view which a native people may adopt
is the calendric, and here I can only speak very tentatively.
I myself have so little appreciation of time that I omitted
almost entirely to enquire into this aspect of Iatmul culture.
My impression is, however, that among the Iatmul there is
very little emphasis upon this aspect of behaviour. The most
important men, with whom it rested to decide when a given
ceremony should occur, gave me occasionally rather inarticulate statements about this periodicity of *wagan* and *mwai*
ceremonies, but though I once or twice tried to get this
periodicity clear, I never succeeded: my informants seemed
to vary in their opinions. Even on the subject of the lunar
calendar informants were but little interested and obviously
ignorant. The difficulty here was that the calendar is not
simply lunar, but is also controlled by the water level.
Nominally the year consists of *twelve* moons, of which five
are moons of high water and five are moons of low water.
Between each of these groups of five moons, there is an intermediate moon. But inasmuch as the rising and falling of the
river is very irregular, my informants were generally doubtful
as to what month it was at any given time. The astronomical
year of 365 days contains of course approximately *thirteen*
lunar months, so that with their theory of twelve moons in
the year, the Iatmul could never be precise in the identification of the moons in their calendar.

In general the occupations of the men seemed to be but
little regulated by the time of day, and the performance of
even the more important ceremonies is liable to be postponed
in the most erratic manner. Ceremonies which are to be
performed at night usually take place when the moon is full,
but they are not infrequently postponed from one full moon
to another, and sometimes this postponement leads to a dance
being held after the water has risen and the dancing ground

is flooded. In such cases a platform is constructed on canoes for the use of the dancers.

I have stated that ethos and eidos are an expression of the affective and cognitive aspects of personality, but it is worth stressing that other aspects of the personality may also be standardised, and we may here very briefly consider some of these aspects:

1. Apollonian and Dionysian. Dr Benedict[1] has described two extreme poles of a possible variation in personality, and has shown that these extremes may be standardised in culture. It is not perfectly clear, however, how the syndromes which she has called Apollonian and Dionysian are related to ethos and eidos. My own impression is that in Apollonian personality we have a standardisation which might occur with many types of organisation of the emotions and sentiments. But until the phenomena of dissociation have been properly related to other psychological phenomena, and especially until we have some idea of what we mean by consciousness, it will not be possible to define these poles of variation more precisely.

2. Tempo. It is a common impression of those who visit foreign countries that the natives are either faster or slower, brighter or duller in their reactions than the members of the observer's own community. This impression is no doubt due to some form of cultural standardisation of the personalities concerned, and should be investigated. We do not know, for example, whether reaction time can be affected by culture.

3. Perseveration. In some cultures it appears that the individuals take great precautions to avoid receiving or giving each other sudden emotional shocks. Miss Lindgren tells me that among the Mongols a man who comes with exciting news, even though his news be of a nature to demand hasty action, will not impart it till some time has elapsed. It is likely that cultures vary considerably in the flexibility which they demand of the individual, and it would be important to know something of the position of an over-flexible individual in a culture which expects a high degree of persevera-

[1] *Patterns of Culture*, 1935.

tion, and the fate of an emotionally stiff individual in a culture which demands flexibility.

4. Lastly we may remember that this book has been devoted solely to a synchronic view of Iatmul culture, and it is possible that individuals and cultures vary in the extent to which they see the world either as a product of the past or as a mechanism working in the present.

At present it is not profitable to do more than suggest the existence of these other types of standardisation of personality, but it is possible that a study of these psychological phenomena in their cultural settings will contribute something to our understanding of the phenomena themselves; and that cultural anthropology has important contributions to make to the subject of individual psychology. Advances in either of these subjects will always contribute to advances in the other, and the concept of cultural standardisation of personality is a link between them.

CHAPTER XVI

Epilogue 1936

THE writing of this book has been an experiment, or rather a series of experiments, in methods of thinking about anthropological material, and it remains to report upon how I was led to carry out these experiments, to evaluate the techniques which I devised, and to stress what I regard as my most important results.

My field work was scrappy and disconnected—perhaps more so than that of other anthropologists. After all, we set out to do the impossible, to collect an exceedingly complex and entirely foreign culture in a few months; and every sincere anthropologist when he comes back to England discovers shocking gaps in his field work. But my case was somewhat worse. The average anthropologist has some definite interest in some one aspect of the culture he is studying, whether it be in historical reconstruction, in material culture, in economics, or in functional analysis, and he will at least collect an adequate supply of material for a book permeated by his particular point of view. But I had no such guiding interest when I was in the field; I was (and still am) sceptical of historical reconstructions; I could not (and still cannot) see that orthodox functional analysis was likely to lead anywhere; and lastly my own theoretical approaches (*Oceania*, 1932, pp. 484 ff.) proved too vague to be of any use in the field.

I did not clearly see any reason why I should enquire into one matter rather than another. If an informant told me a tale of sorcery and murder, I did not know what question to ask next—and this not so much from lack of training as from excess of scepticism. In general therefore, apart from a few standard procedures such as the collection of genealogies and kinship terminology, I either let my informants run on freely from subject to subject, or asked the first question which came into my head. Occasionally I made an informant go back to some previous subject of conversation, but I should have found it hard to give theoretical reasons for such special attention to certain subjects.

This method, or lack of it, was wasteful, but it has its compensations. I know, for example, how great a value the natives set upon their enormous system of totemic names; and I know this, not from some colourless statement, but from the curious experience of writing down thousands, literally, of names at my informants' bidding; and I know it too from the more bitter experience of finding my note-books full of names when I searched them for information about preferred types.

The arrival of a part of the manuscript of Dr Benedict's *Patterns of Culture*, an event already mentioned in the Foreword, together with conversations with Drs Fortune and Margaret Mead, gave me a vague clue to what I wanted to do in anthropology, and in the last three months before I left New Guinea I tried to follow this up. I realised then the importance of method in field work, and endeavoured in the last part of my time to make up for some of the deficiencies in my notes on the most ordinary subjects. I recognised the significance of the contrast in ethos between the sexes, but I did not devise any special methods of enquiry which might have been relevant to an ethological approach. It was only after I had returned to England that I realised the importance of observing the reactions of one sex to the behaviour of the other, or of collecting native statements about preferred types. When an informant told me that Woli-ndambwi had a big nose (cf. p. 162) I wrote his statement down, but with no idea that this detail of the culture might be of any particular interest.

I am here stressing the lack of method in my field work in order to satisfy those who may say that I have "selected my facts to fit my theories". Some selection has, of course, occurred in the process of creaming my note-books, but this has been done in England. The actual noting of facts was done at random before I even dreamt of ethos, eidos and schismogenesis, and I emphasise these considerations now because they will never be applicable to my future work when I hope to select facts which shall be relevant to my problems.

When I came to the task of fitting my observations together into a consecutive account, I was faced with a mass of the most diverse and disconnected material. I had for example a number of stories of sorcery and retribution, but to not one of them was appended a systematic study of the relevant facts of kinship, technique, emotional attitude, etc. I had collected, of course, not isolated facts, but facts in little bunches: the facts in some bunches were grouped on a chronological basis, and in other bunches they were grouped on a structural basis, and so on. No one system of organisation ran through the material, but in general my groups of facts had been put together by my informants, so that the systems of grouping were based upon native rather than scientific thought. Out of this material I had to construct a picture of Iatmul culture. If in the circumstances I have managed to demonstrate some coherence in the culture, this achievement is the best testimony to the usefulness of the methods of analysis.

An account of the various steps which have led to the construction of this book will serve as a summary of my suggestions, and the narration of the errors, into which I fell by the way, may be of use to any who care to follow.

One detail of the Palimbai *naven* had struck me with great force. I had previously seen transvestite women, proud of their male ornaments, and had even published an account of the Mindimbit *naven* ceremonies, but I had not before seen the transvestite *wau*. I had never realised that he was a figure of fun. My whole mental picture of *naven* had been wrong, and wrong because, though I had been told what was done, I had no idea of the emotional aspects of the behaviour.

Though I did not know what it meant, I knew that the *wau*'s buffooning had altered my whole conception of *naven*, and, if this was so, the contrast between the bedraggled transvesticism of the men and the proud wearing of homicidal ornaments by the women must somehow contain an important clue to Iatmul culture. The change in my way of thinking had arisen from the addition of emotional emphasis to what was originally a purely formal picture, and so I came to believe that ethos was the *thing* that mattered.

Much later, after I had returned to England, I found from my photographs that women, when they are decorated for public ceremonial, wear ornaments such as are usually worn only by men, and this discovery led me to consider the analogy of the fashionable horsewomen and to develop the theory of Iatmul transvesticism which I have put forward in Chapter xiv.

How sound this theory may be I can hardly judge, but it is put forward in all seriousness, and I have a special affection for it since, though now the theory stands as a rather unimportant detail in the book, it was from this detail that the whole synthesis grew. At the International Congress in the summer of 1934, I read a paper to the Sociology Section on my theory of Iatmul transvesticism. In it, I gave sketches of the *naven* ceremonial and of the ethos of each sex in Iatmul culture. I felt that I had accounted rather satisfactorily for the curious details of *naven*, and I was still sure that ethos was the thing that mattered.

After the Congress, I started to rewrite this small paper on transvesticism, thinking that it was worth publishing, and that I could afterwards use the paper as the skeleton of a book on Iatmul culture. But the paper grew by the addition first of one method of approach and then of another, until it has become the present book, and now its purpose is no longer to put forward a theory of Iatmul transvesticism, but to suggest methods of thinking about anthropological problems.

The first difficulty that I encountered was over the fact that Iatmul men wear aprons and Iatmul women wear skirts. I felt that my ethological theory was a fairly satisfactory description of the ceremonial transvesticism, but I could not account for the simple circumstance that on more ordinary occasions special clothes were worn by each sex; and this differentiation was clearly necessary if transvesticism was to occur in the culture. But the problem did not worry me seriously, and I dismissed the matter. It was a rule of the culture, a "formulation", that individuals of different sex should wear different clothes. Indeed, I dismissed the matter so effectively that I have nowhere in the book stated the obvious fact that the differentiation of clothing according to sex is one of the factors which "promote" transvesticism.

The matter became more serious when I asked why it should be the mother's brother who performs those antics and not the father. Again I took refuge in the terms "formulation" and "structure". The kinship system, "built up" of formulations, was part of the structure of society, and the *naven* was "built upon" this structure. It appeared, then, that there was another "thing" called "structure" which mattered in the culture.

While ethos appeared to me to consist of preferred types and behaviour which was expressive of emotion, structure consisted of the kinship system and other "formulations". If a man scolded his wife, his behaviour was ethos; but if he married his father's sister's daughter, it was structure. I even went so far as to think of the structure as a network of channels which guided the ethos, and were shaped by it (cf. footnote, p. 121). Looking back now, it is almost incredible that I ever thought along such lines or used such metaphors, and I find it quite difficult to write of my early theories without caricature.

It took quite a long time to get away from these fallacies, and I had advanced in other directions before I escaped. I added what I used to call "pragmatic function" to my list of the subdivisions of culture. This was a mixture of the "satisfaction of the needs of individuals" and the "integration of society". The confusion between these two sorts of social function was a result of muddling two almost unrelated facts: that allegiance is needed by the *wau*, and that it also serves to integrate society. I still regarded ethos, structure, and pragmatic function as categories into which culture could be subdivided, and I had even started to write the last chapter of the book before I escaped from this morass.

In this last chapter I intended to examine the inter-relations between these different subdivisions of culture, and I found that I had given no clear criterion for discriminating the elements of culture which I would pigeon-hole as ethos from those which I would pigeon-hole as structure or pragmatic function. I began to doubt the validity of my categories, and performed an experiment. I chose three bits of culture: (*a*) a *wau* giving food to a *laua*; a pragmatic bit, (*b*) a man scolding his wife; an ethological bit, and (*c*) a man marrying his father's sister's daughter; a structural bit. Then I drew a lattice of nine squares on a large piece of paper, three rows of squares with three squares in each row. I labelled the horizontal rows with my bits of culture and the vertical columns with my categories. Then I forced myself to see each bit as conceivably belonging to each category. I found that it could be done.

I found that I could think of each bit of culture structurally; I could see it as in accordance with a consistent set of rules[1] or formulations. Equally, I could see each bit as "pragmatic", either as satisfying the needs of individuals or as contributing to the integration of society. Again, I could see each bit ethologically, as an expression of emotion.

This experiment may seem puerile, but to me it was very important, and I have recounted it at length because there may be some among my readers who tend to regard such concepts as "structure" as concrete parts which "interact" in culture, and who find, as I did, a difficulty in thinking of these concepts as labels merely for points of view adopted either by the scientist or by the natives. It is instructive too to perform the same experiment with such concepts as economics, kinship and land tenure, and even religion, language, and "sexual life" do not stand too surely as categories of behaviour, but tend to resolve themselves into labels for points of view from which all behaviour may be seen.

The matter may be re-phrased, either in terms of points of view adopted by the scientist, or in terms of aspects of the behaviour. We must expect to find that every piece of behaviour has its ethological, structural and sociological significance.[2]

---

[1] This legalistic phrasing of cultural structure I have since dropped, and have substituted the term *premises* for rules or formulations, emphasising the *consistent* nature of cultural structure and ignoring the question of whether it is enforced, and whether it is articulately formulated by the natives.

[2] I do not claim to have detected the fallacy of "misplaced concrete-

This meant that I had only to keep clearly before me the conviction that ethos, structure, etc. were merely points of view or aspects of the culture, and to look for each aspect in every piece of behaviour and in every native statement. But this was still difficult. In the first place, the habit of thought which attributes concreteness to aspects of phenomena is one which dies hard. The fallacy which Whitehead has pilloried has been an important principle or motif of European eidos, certainly since the days of Greek philosophy. It has taken me over a year to drop the habit even partially, and I fear that many passages in the book may be still more or less infected with it, in spite of drastic revision.

Another difficulty was that of keeping my points of view distinct and separate. Constantly I wandered from the point of view which was appropriate in the chapter on which I was working and found, for example, that I had inserted a paragraph of structural phrasings into a chapter on ethos. One instance of this is worth narrating: I had been much impressed by the tone of the debate on the profanation of the junior ceremonial house (cf. p. 100) and from my enquiries about this debate the fact had emerged that there are no serious sanctions inside an initiatory grade. This I recognised as a fact of prime importance, and I made a note that this lack of internal sanctions should be inserted into the chapter describing ethos in the ceremonial house. Later I wrote several pages on the matter for insertion there, and it was not until I re-read what I had written that I found that these pages were nothing but sociology. They then became the nucleus of the second half of the sociological chapter.

This example will serve to show how misplaced concreteness tends to confuse the separable aspects. My sequence of thought, though inarticulate, had run something like this: the tone of this debate is striking; the debate is "pure ethos"; therefore the generalisation which I got from that debate is ethos.

I was bothered, too, by the feeling that if the natives them-

ness" independently of Whitehead, and should perhaps never have succeeded in disentangling the matter if conversations with C. H. Waddington had not planted the seeds of Whitehead's philosophy in my mind.

selves discriminated, as the natives of Western Europe do, between economics and law, and regarded these as subdivisions rather than as aspects of culture, I ought to accept their view of the matter and adopt their categories. Later, when I was working on Iatmul eidos it occurred to me that we ought to look for a standardisation in the various cultures of the world of those very methods of thought which I had myself been advocating: and when I realised that people might think sometimes structurally and sometimes in terms of economics and sometimes in terms of ethos or sociology, then I understood that I need worry no more about native ideas of the subdivision of culture. I was justified in expecting every aspect to be represented in every bit of behaviour, and the fact that the natives were conscious only of one of these aspects in a given context was a point which might be significant in an eidological examination of the culture.

For me, the escape from the fallacy of misplaced concreteness was an enormous advance, and I hope that others may find that the account of my adventures helps them to see their problems more clearly and simply. From then on, the construction of the system of abstractions which I have put forward was a comparatively easy matter. But three other steps in the process may be mentioned, if only to give honour where honour is due:

The separation of sociology from my jumble of "pragmatic functions". This was due to Professor Radcliffe-Brown. I gave a paper in Chicago outlining my system of abstractions, and Professor Radcliffe-Brown observed that I used the term "structure" in a sense different from his; that he used it to refer to the structure of *society*, while I was using it for what he then proposed to call "cultural structure". Following this up, I resolved my "pragmatic function" into two separate abstractions: "sociology" in the strict sense of the term and what I have called "motivation" or the "expression of ethos in behaviour".

The isolation of eidos from ethos. This resulted from the awkwardness of my use of the term "logic" in the definition of cultural structure. I found that a footnote was necessary to explain that we must expect to find different sorts of "logic" in different cultures, and in writing this footnote I realised that I ought to describe the logic of Iatmul culture as fully as I could, and that this was an important aspect of the culture. "Ethos", in my old phrasing, was subdivisible then into two major abstractions, ethos and eidos, which were related to each other as the affective aspects of behaviour are related to the cognitive. I therefore dropped the old, wider use of the term "ethos" to cover all those pervasive characteristics of a culture which can be ascribed to the standardisation of the individuals, and now use Dr Benedict's term, *configuration*, in this wide sense. Dr Benedict has agreed with me in conversation that, in her original use of this term, she intended it to include the standardisation of many different aspects of personality, and it is convenient to have a general term for the combination of all these aspects of culture.

It is unfortunate that, although the awkward word "logic" led to the concept of eidos, the working out of this concept has contributed virtually nothing to mitigate the awkwardness of "logic". This I think is perhaps the weakest point in the exposition of my system. I am reasonably satisfied that cultural structure is an important and isolable aspect of behaviour, but the exact phrasing of the sort of consistency characteristic of culture when seen from this viewpoint has still eluded me.

Closely connected with this difficulty is that of discriminating the affective from the cognitive aspects of personality. I am aware that psychologists are inclined nowadays to look askance at the terms affective and cognitive, but I still hope that if we first grant that these terms are aspects and not categories of behaviour, and then go on to compare the ethos and eidos of various cultures, we may in the end arrive at a better understanding of thought and emotion.

Schismogenesis. This concept has developed very slowly from conversations with Drs Margaret Mead and Fortune. Dr Mead contributed an idea of the very first importance, that of complementary ethos. But for a long time after I returned to England I still thought of the phenomena in static terms, as types of ethos or types of personality. I allowed myself to be influenced unduly by the conclusions of typological psychology. Gradually I realised that if there existed this complementary relationship between the ethoses of the two sexes in Iatmul culture, it was evident from the special nature of this relationship that each ethos must have some formative or directive action upon the other. From such concrete phrasings of the matter the concept of complementary schismogenesis developed.

Later, out of a conversation about European politics with Alan Barlow, I developed the idea of symmetrical schismogenesis. The chapter on these processes is a result of combining these two lines of thought and leavening the mixture with a little Hegelian dialectic.

Such then have been the steps which have led me to the isolation of five major points of view for the study of the behaviour of human beings in society—structural (and eidological), emotional (and ethological), economic, developmental and sociological. I have given no samples of the developmental and economic methods, but I have tried to give a sketch of Iatmul culture from each of the other points of view. The result has been that in each chapter, according to the method adopted, different parts of the culture have been set side by side. In the structural chapter, the most diverse contexts of the culture have been set side by side to demonstrate the identifications relevant to the *naven* behaviour of the mother's brother. In the sociological chapter we have seen the *naven* ceremony set beside the marriage system as part of a great mechanism for the integration of the society. In the ethological chapters, characteristic expressions of Iatmul emotion have been collected together from the contexts of everyday life, head-hunting and initiation, and from this collection we were able to go on to attribute specific emotional value to the behaviour of the various relatives in *naven*. Finally in the chapter on Iatmul eidos, I have set side by side the patterns of Iatmul totemism, initiatory grades and even flute music to give a picture of the ways of thought of the Iatmul. In fact, whatever our method, our material is the same and includes the whole ordered diversity of Iatmul behaviour.

But though the material for our studies is the same what-

ever aspect of the culture we intend to study, it is by no means
immaterial in what order we investigate the different aspects.
In attributing emotional value to the details of *naven* behaviour, I have insisted that any such attribution is dangerous
until the ethos of the whole culture has been studied; and
therefore the general picture of Iatmul ethos necessarily pre-
cedes the chapter in which I have tried to analyse the
emotions of the various relatives in *naven*—and even with
this precaution, this chapter is perhaps the least satisfactory
in the book.

In the study of the logic of Iatmul culture I have adopted
the opposite procedure. I first outlined the premises upon
which the *naven* ceremony is based and only later gave a
general picture of Iatmul eidos. The question must therefore
be asked: why, if it is not justifiable to guess at motives before
the whole ethos has been studied, is it permissible to attribute
behaviour to specific premises before the eidos of the culture
has been sketched?

This question is, I suspect, one of considerable importance,
and its answer may well reflect upon the relation between
the affective and cognitive aspects of the personality. At
present I can only suggest two possible answers: the order
of procedure which I have been forced to adopt may be a
result either of my own psychology or of that of my informants.
I have stated that in my own conscious mental processes,
the structural and logical aspects of behaviour are more
clearly seen than the emotional, and I believe that the same
is true of the Iatmul men. If this observation be correct,
then in attributing emotional value to details of behaviour
I was speculating about the more unconscious processes and
therefore needed the supporting material which could only
be provided by a preliminary study of Iatmul ethos. But in
the case of the logical aspects of behaviour I could hope for
some degree of articulateness in my informants and could
proceed directly to the statement of consistency between
different details of behaviour.

Moreover, a greater articulateness about the structural
aspects of behaviour is characteristic not only of the natives
but also of Europeans. The English language—and perhaps

all languages—is not adapted to the critical description of emotion. While a syllogism can be stated in stark simplicity, it is almost impossible to make any simple statement about emotions. The best we can do is to sketch an ethos or a personality in a diffuse journalistic or artistic style and then sum up that sketch in a few words whose significance is fixed by the sketch. Thus after giving a sketch of Iatmul ethos, I labelled that ethos as "proud" and "schizothyme". These words are capable of a hundred different meanings, but the sense in which I have used them is defined to some extent by my sketch of Iatmul ethos. Failing such a preliminary sketch, our attribution of emotional value to behaviour can only be guided by general and probably fallacious assumptions about human nature, assumptions which are so general that they blur all differentiation of sex, temperament and culture.

In this final chapter we are trying, prematurely perhaps, to get some general view of the different methods of approach which have been sampled in the course of the book, and one point is of special interest. If, whatever method we adopt, our material is the same and the only difference between one method and another lies in the arrangement of the material, then it is relevant to enquire whether any subdivision of the culture or any other systematic phenomenon has proved so fundamental as to survive these re-shufflings of the details of behaviour and to reappear in recognisable form in each of our pictures.

The orthodox "subdivision" of culture into such institutions as marriage, kinship, initiation, religion, etc. has entirely disappeared; but one phenomenon has persisted through all the re-shufflings and has been described in different terms in each picture. In the chapter on ethological contrast and competition, actually the last chapter to be written, I stated that though the concept of schismogenesis had developed out of the study of ethos, "we should be prepared to study schismogenesis from all the points of view—structural, ethological and sociological—which I have advocated". When I wrote this sentence I intended to postpone the study of the sociological and structural aspects of schismogenesis to some future work. I was aware that the Marxian historians·had occupied themselves with the economic aspects of the process and I had deduced from this that we must expect structural and sociological aspects, but I had no idea what the process might look like from these points of view. It was therefore a surprise to me to find, on re-reading my manuscript, that I had actually devoted several pages of my chapters on eidos and sociology to the two forms of schismogenesis; and the surprise was the greater inasmuch as the concept of symmetrical schismogenesis had been a very late addition to my theory.

The last pages of the chapter on sociology (pp. 106 *et seq.*) are devoted to a discussion of the various patterns of fission which occur in human societies. I here stressed the contrast between the fission of Iatmul communities, which gives rise to daughter groups with cultural norms similar to those of the parent society, and the fission of European communities which so often produces daughter groups with cultural norms divergent from those of the parent. I noted also that the production of daughter groups with similar norms was associated with a peripherally oriented system of sanctions, whereas the production of divergent daughter groups was associated with a centripetal or hierarchic system.

These two types of fission are simply describable in terms of symmetrical and complementary schismogenesis. I stated that in the integration of Iatmul society the weakest links are those between affinal relatives and that it is these links which are broken when the community divides, while the links based upon patrilineal relationships persist. The "plane of cleavage" of this society is parallel to the patrilineal but cuts across the matrilineal and affinal bonds.

We have seen that in this society the relationships between father and son and between elder brother and younger brother are complementary, while that between brothers-in-law is symmetrical. Thus it is the symmetrical relationships which are chiefly broken in fission, and we may reasonably suspect that the fission itself is in part a result of symmetrical schismogenesis.

We found that one of the sociological functions of *naven* is the strengthening of these affinal links, but it was not perfectly clear (cf. p. 96) why for this purpose it should be relationship between classificatory *wau* and *laua* which is stressed rather than that between classificatory brothers-in-law. It would seem to the European mind to be the more natural to stress the latter relationship. But in the light of the schismogenic position it is reasonable to suppose that the *wau-laua* relationship with its accompanying difference in age between the persons is more consistently complementary in its behaviour patterns and therefore more adapted to function as a bond in a society which seemingly has some difficulty in controlling symmetrical schismogenesis. Moreover, it is worth noting that the actual behaviour of the *wau* in *naven* may be described as an insistence upon the complementary aspects of his relationship with his *laua*, at the expense of the symmetrical aspects. He stresses his position as "mother" and "wife" of his *laua*, thus denying the symmetrical aspect of the relationship connected with his position as "brother-in-law".

It is possible that this insistence on the complementary patterns in the *wau-laua* relationship is a case of the control of a symmetrical schismogenesis by admixture of complementary patterns of behaviour (cf. p. 193). But, since the final fission of the community still follows a plane of cleavage which cuts across these matrilineal and affinal bonds, we must suppose that this attempt to control the symmetrical schismogenesis is not entirely successful.

In contrast to the fission of Iatmul communities by symmetrical schismogenesis, we find that, in European communities, the daughter groups generally separate from the parent in revolt against the elaborate centripetal hierarchies—legal, religious and military—which are a characteristic feature of the integration of these communities. These hierarchies are all of them built up of series of complementary relationships, and it is therefore reasonable to suppose that the fissions are either wholly or in part a result of complementary schismogenesis.

More than this we cannot at present say; and it is likely that in attributing the fission of Iatmul communities to symmetrical schismogenesis and that of European communities to complementary schismogenesis I have been more sweeping in my generalisations than is strictly justified. These sweeping statements will, however, serve as sample of the generalisations which might be carefully documented by a study of schismogenesis in its sociological aspects.

While in the sociological chapter, schismogenesis appeared in the guise of a classification of types of social fission, the same process was represented in the chapter on Iatmul eidos as a classification of types of dualism (pp. 237–247). I was dissatisfied with my phrasing of "direct" and "diagonal" dualism and had found considerable difficulty in explaining these concepts to others; but I did not see my way to a re-phrasing of the matter until I realised that for the terms "direct" and "diagonal" I could substitute *complementary* and *symmetrical* in the senses in which these terms had been used in the description of schismogenic relationships.

This equivalence is clear from a consideration of the list of relationships which I have classed respectively as "direct" and "diagonal". The "direct" relationships are those between elder brother and younger brother, brother and sister, and man and wife; and all of these are typical complementary relationships. The "diagonal" relationships are those between brothers-in-law (especially brothers-in-law who have exchanged sisters) and between *kaishi* (counterparts who have exchanged presentations of valuables, or persons whose children have inter-married). Both of these relationships are symmetrical.

In the "direct" relationships the importance of the complementary patterning is evident from the insistence in the lists of totemic names on difference of age between siblings of the same sex. In the case of siblings of opposite sex, it is significant that there is no suggestion of a difference in age, "the difference in sex being apparently analogous to difference in age" (p. 243). We may now guess at the nature of this analogy: these two sorts of difference are alike in that each introduces a basis for complementary patterns into the relationship.

In the relationships between pairs of clans and between the two totemic moieties we observed that both forms of dualism are recognisable. While the clans are nominally grouped together in pairs as "elder" and "younger", the members of the clans also devote a great deal of energy to symmetrical competition in pride of ancestry and, as a result, the types of dualism are tangled together. We may suggest, very tentatively, that the statement of elder and younger brotherhood between related clans perhaps mitigates the symmetrical competition between them. But the matter is not clear and needs more study of the actual behaviour; especially we need to know in what contexts the members of a pair of clans insist upon their brotherhood and in what contexts they compete (cf. footnote, p. 242). Without more complete information we must disregard these groupings as sources of evidence for the equivalence between the types of dualism and the types of schismogenic relationship.

This equivalence is, however, clearly demonstrable in the elaborate systems of alternation of initiatory grades and of brothers. I described this alternation as due to "a combination of the two types of dualist thinking, an intrusion of diagonal thought into contexts where direct identification and the principle of seniority were already at work" (p. 247). In terms of schismogenic patterns we may now re-state this combination of the two forms of dualism. The relationship between any two consecutive grades is complementary and is marked by such steep contrasts as occur in initiation, where the novice is the "wife" of the initiators. But the relationship between the initiatory moieties is symmetrical. Rivalry and combat are culturally expected between the two groups and, in fact, this relationship provided the most clearly developed of our examples of symmetrical schismogenesis.

As to the mechanism which determines this building up of alternating series, I have assumed (p. 244) that the alternating pattern is itself a motif of Iatmul eidos and that therefore this pattern reappears in whatever contexts are suitable. But it is not clear that this assumption is justified or necessary. It is possible that not the whole pattern but only the two types of dualism are carried in the eidos. Thinking for a moment in diachronic terms, we can see that it is unlikely that any of these complex alternating systems has been introduced entire by culture contact. We therefore have to consider how the first of these systems can have developed; only after this will it be permissible to think of the alternating pattern as "carried by the eidos", i.e. to suppose that the other alternating systems are built up on the model of the first. Moreover, I know of no native terminology to describe alternation, and in the absence of such terminology it is likely that the alternating systems are formed afresh in each context. It is possible that the only standardisation provided by the eidos is that of the two types of dualism. These the native language expresses adequately by means of kinship terminology.

If this be so, then we must see the development of alternating systems in Iatmul culture and their absence in our own culture as a function of the fact that among the Iatmul *both* complementary and symmetrical patterns are thought of in dual terms, while in Europe though we regard complementary patterns either as dual or as arranged in hierarchies, we do not think of patterns of rivalry and competition as necessarily dual. Rivalry and competition in our communities are thought of as occurring between any number of persons with no presumption that the resulting system will be patterned upon any sort of bilateral symmetry. Only if both types of relationship are habitually thought of in dual terms is it likely that alternating hierarchies of the Iatmul type will be developed.

Another line of enquiry may be mentioned, although no answer can be given to the questions which it raises. As in the case of contrasting ethos we were led to ask how the appropriate system of emotional attitudes is inculcated in each individual, so here we may ask how the two methods of dualist thought are inculcated. To this question I will only reply with the awkwardly phrased suggestion that such habits of thought are a result of observation of, and conversation about, patterns of relationship.

But the question itself is of some academic interest in the present context because it illustrates perhaps better than any other the difference between ethos and eidos; and therefore the examination of its terms may lead to a proper formulation of the difference between affective and cognitive aspects of personality.

In the ethological problem we had to decide how an

individual is moulded so that he himself adopts a certain ethos, a certain system of emotional attitudes in his contacts with other people and the outside world. In behaviourist terminology, we had to decide how a certain *system of responses* is conditioned. But in the present problem we have to decide how the individual is trained to discriminate and classify relationships between persons and between groups. Thus we are here concerned, not with systems of responses, but with the system of classification which the individual uses upon the stimuli.

We may therefore say that every series of events of the type which we may call a stimulus-response has both affective and cognitive aspects; and the aspect which we shall study is determined by the way in which we work. If we take a given stimulus-response and consider all the details of the *response* and how they are linked together, and if we then go on to analyse the composition of other sets of linked details of response, we shall be studying the organisation of responses and shall arrive at a picture of the affective aspects of the personality. If, on the other hand, we start from a given stimulus-response and compare this with other similar series of events to obtain first a list of *stimuli* which evoke a particular detail of response and then a classification of stimuli according to the responses which they evoke, we shall finally arrive at a picture of the cognitive aspects of the personality.

In these procedures, it may be noted that strictly speaking we arrive not at pictures of the individual but at pictures of the events in which the individual is involved. This inconsistency disappears when we realise that the term "personality" refers not to the isolated individual but to the *individual in the world*.

This definition of the terms "affective" and "cognitive" may be expressed diagrammatically (Figs. 5 and 6). When we find that an individual responds with reaction, *A*, to the stimuli 1, 3 and 5, we may say that he discriminates these stimuli from 2, 4 and 6, and classifies them together on the basis of some quality which they possess in common; and we might go on from this to a more elaborate study of the whole system of discriminations and classifications which are characteristic of that individual. Such a study would be a study of the cognitive aspects of his personality.

At an early stage in this book (p. 25) I spoke of "syllogisms" and "logic" in cultural structure; and it is of some interest to consider how this would have to be re-phrased if future work should show this definition of cognitive aspects of personality to be valuable. Briefly we should have to phrase

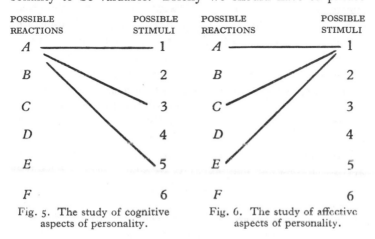

POSSIBLE REACTIONS   POSSIBLE STIMULI   POSSIBLE REACTIONS   POSSIBLE STIMULI

Fig. 5. The study of cognitive aspects of personality.

Fig. 6. The study of affective aspects of personality.

the matter not in terms of logic but in terms of classifications of behaviour. Instead of saying: "The individual exhibits reaction, *A*, in response to stimuli 1 and 3; stimulus 5 resembles (or is identified with) 1 and 3; therefore the individual exhibits reaction *A* in response to stimulus 5", we should rather say: "Stimuli 1, 3 and 5 are classified together". From this we should proceed directly to the study of the system of classification, the eidos of the culture.

On the affective side the matter is not quite so simple. If we find that the individual reacts to stimulus, 1, with the simultaneous reactions, *A*, *C* and *E*, we may classify these reactions together as a syndrome of symptoms of some emotion. From this we may go on to a classification of all the series of linked responses exhibited by the individual and so get a general description of the affective sides of his personality. But there are difficulties in this view and it is possible that there are other mechanisms which link responses together, besides those systems which we may describe as "emotions". For the present, however, I know of no phrasing of the matter which so nearly translates the term "emotion" into behaviouristic language.

Such a phrasing of affect would I think have had but little effect upon my general statement of Iatmul ethos. It is true that the description which I gave of Iatmul ethos is very far from a systematic account of sets of linked responses, but this shortcoming is due, not to a weakness in my theoretical concepts, but to the merely practical difficulty of describing human behaviour in a critical and comprehensive manner. When I stated that the "tone" of the men's behaviour in the initiation ceremonies was expressive of harshness and irresponsibility rather than of asceticism, I meant that the actions performed by them, the washing of the novice, etc., were *accompanied* by other details of behaviour so that the whole picture was one of harshness. Until we devise techniques for the proper recording and analysis of human posture, gesture, intonation, laughter, etc. we shall have to be content with journalistic sketches of the "tone" of behaviour.

Some modification would, however, have been necessary in the chapter entitled "The Expression of Ethos in *Naven*". The term "motivation" would have disappeared entirely, but it is not quite clear what would have taken its place. I think that ideally the chapter should have consisted of an analysis of all the syndromes of response exhibited by the *wau*. These syndromes would then have been compared and equated with syndromes of response collected in the other contexts of the culture, and finally the chapter would have developed into a synthesis of affective and cognitive studies to show that the stimuli of the *naven* situation were such as might be expected to evoke these syndromes.

From this contemplation of the elaborate analysis which ideally ought to have been done, we may turn to what actually was done and examine more carefully the differences between the various points of view which I adopted in the various chapters. We have seen that schismogenesis is recognisable in three different aspects of the culture, in ethos, in eidos and in sociology; and we may profitably consider what

distortions were introduced into the picture of this diachronic process according to the various methods of its
delineation.

In the first place, it is certain that all the representations
of schismogenesis should have had one failing in common.
They should have given us no indication that schismogenesis
is a diachronic process, a process of change. Each method
was nominally strictly synchronic and each should have given
only a synchronic "section" of schismogenesis in which the
process of change would have been seen as stationary in a
single instant of time.

Actually, the sociological view of schismogenesis, owing
to a lack of stringency in my exposition, showed us the two
forms of social fission, and these must I think be regarded
as diachronic processes. Strictly in the sociological chapter
I should have confined myself to discriminating between
peripheral and centripetal systems of integration and should
not have wandered from this classification to consider the
results of fission in the two cases. It is conceivable that even
from this strictly limited classification we might have arrived
at the statement that peripheral integration is chiefly dependent upon symmetrical links between individuals, while
centripetal integration is chiefly dependent upon complementary links. But I did not grasp this at this stage, probably
because my attention was fixed too rigidly upon the major
unit, the society as a whole.

In the description of Iatmul ethos, however, the picture
which we drew of schismogenesis was strictly synchronic.
We saw ethological contrast as statically existing in the cul
ture.[1] Indeed from the labelling of the ethological contrast,
first as "complementary" and then in terms of Kretschmer's
typology, to the final realisation that each ethos must have a
formative effect on the other, was a very long step. The diachronic aspects of schismogenesis did not come into view
until I attacked the question of the processes of moulding
of individuals into the complementary patterns, until, in fact,

[1] The analysis of the schismogenic processes in initiation was a late
insertion into the description of ethos and was written *after* I had discriminated the two types of schismogenesis.

I left ethology and began to think in terms of developmental psychology.

The view of schismogenesis provided by the sketch of Iatmul eidos was equally limited in its scope and I think it unlikely that I should ever, from this aspect alone, have described "direct" and "diagonal" dualism in terms of complementary and symmetrical patterns of relationship. Here the difficulty in the way of a clear phrasing of the matter was due not merely to the lack of diachronic perspective but also to two other circumstances: first, I had already attached labels to the two forms of dualism and it would have been difficult to get away from the implications of these labels; and secondly, I had realised that it was immaterial to the dualism whether the units of a given pair were identified or discriminated and this (illogically) would have prevented me from inspecting the patterning of the relationships.

Thus each of our separate methods involves distortions and gives but a partial view of the phenomena, but I do not think that for this reason the methods should be discarded. Against the incompleteness of each isolated aspect of the phenomena, we may set the increased simplicity which such isolation provides and the advantages of being able to state our problems and generalisations in such clear terms that we can at once look for the relevant facts. Moreover, it may be remembered that the architect who draws first a plan and after that an elevation of a building is not thereby prevented from finally drawing a picture of the building in perspective. After we have looked at our sketches of isolated aspects of the culture, it will still be possible to go on to a synthesis of these aspects, a synthesis which will involve its own types of distortion but which will be the more complete because in collecting our facts we shall have been guided by several different types of relevance.

Finally, it remains for me to state what value I attach to the facts, the theories, and the methods, which I have put forward. It is clear that I have contributed but little to our store of anthropological facts and that the information about Iatmul culture which I have used in the various chapters does no more than illustrate my methods. Even for purposes of illustration my supply of facts is meagre, and I certainly cannot claim that my facts have demonstrated the truth of any theory.

This would be a serious confession of weakness, and indeed the book would have no value, were it not for another shortcoming which in a sense cancels out the first. To have put forward, unsupported by a solid backing of fact, theories and hypotheses which were new, would have been criminal, but it so happens that none of my theories is in any sense new or strange. They are all to some extent platitudes, which novelists, philosophers, religious leaders, lawyers, the man in the street, and even anthropologists, have reiterated in various forms probably since language was invented. Structure, ethos, and the rest are not new ideas or new theories: they are only new labels for old ways of thought, and indeed only two of my abstractions eidos and schismogenesis—can claim the distinction of receiving even new labels.

I am aware that there are dangers inherent in the use of labels, and that these little pieces of paper are only too liable to hide the things to which they are attached. But though labels must always be used with caution, they are useful things, and the whole of science depends on them. In this case, their use has been to help me to disentangle old ideas and to enable me to think in terms of one aspect of culture at a time instead of jumbling all the aspects together. Our unscientific knowledge of the diverse facets of human nature is prodigious, and only when this knowledge has been set in a scientific framework shall we be able to hope for new ideas and theories.

CHAPTER XVII

Epilogue 1958

THERE is a well-known story about the philosopher Whitehead. His former pupil and famous collaborator, Bertrand Russell, came to visit Harvard and lectured in the large auditorium on quantum theory, always a difficult subject, and at that time a comparatively novel theory. Russell labored to make the matter intelligible to the distinguished audience, many of whom were unversed in the ideas of mathematical physics. When he sat down, Whitehead rose as chairman to thank the speaker. He congratulated Russell on his brilliant exposition "and especially on leaving . . . *unobscured* . . . the vast darkness of the subject."

All science is an attempt to cover with explanatory devices—and thereby to obscure—the vast darkness of the subject. It is a game in which the scientist uses his explanatory principles according to certain rules to see if these principles can be stretched to cover the vast darkness. But the rules of the stretching are rigorous, and the purpose of the whole operation is really to discover what parts of the darkness still remain, uncovered by explanation.

But this game has also a deeper, more philosophic purpose: to learn something about the very nature of explanation, to make clear some part of that most obscure matter—the process of knowing.

In the twenty-one years that have elapsed since the writing of this book, epistemology—that science or philosophy which has for subject matter the phenomena which we call knowledge and explanation—has undergone an almost total change. Preparing the book for republication in 1957 has been a voyage of discovery backwards into a period when the newer ways of thought were only dimly foreshadowed.

Naven was a study of the nature of explanation. The book contains of course details about Iatmul life and culture, but it is not primarily an ethnographic study, a retailing of data for later synthesis by other scientists. Rather, it is an attempt at synthesis, a study of the ways in which data can be fitted together, and the fitting together of data is what I mean by "explanation."

The book is clumsy and awkward, in parts almost unreadable. For this reason: that, when I wrote it, I was trying not only to explain by fitting data together but also to use this explanatory process as an example within which the principles of explanation could be seen and studied.

The book is a weaving of three levels of abstraction. At the most concrete level there are ethnographic data. More abstract is the tentative arranging of data to give various pictures of the culture, and still more abstract is the self-conscious discussion of the procedures by which the pieces of the jigsaw puzzle are put together. The final climax of the book is the discovery, described in the epilogue—and achieved only a few days before the book went to press—of what looks like a truism today: that ethos, eidos, sociology, economics, cultural structure, social structure, and all the rest of these words refer only to scientists' ways of putting the jigsaw puzzle together.

These theoretical concepts have an order of objective reality. They are *really* descriptions of processes of knowing, adopted by scientists, but to suggest that "ethos" or "social structure" has more reality than this is to commit Whitehead's fallacy of misplaced concreteness. The trap or illusion—like so many others—disappears when correct logical typing is achieved. If "ethos," "social structure," "economics," etc., are words in that language which describes how scientists arrange data, then these words cannot be used to "explain" phenomena, nor can there be any "ethological" or "economic" categories of phenomena. People can be influenced, of course, by economic theories or by economic fallacies—or by hunger—but they cannot possibly be influenced by "economics." "Economics" is a *class* of explanations, not itself an explanation of anything.

Once the fallacy has been detected, the way is open for growth of an entirely new science—which has in fact already become basic to modern thought. This new science has as yet no satisfactory name. A part of it is included within what is now called communications theory, a part of it is in cybernetics, a part of it in mathematical logic. The whole is still unnamed and imperfectly envisioned. It is a new structuring of the balance between Nominalism and Realism, a new set of conceptual frames and problems, replacing the premises and problems set by Plato and Aristotle.

One purpose, then, of the present essay is to relate the book to these new ways of thought which were dimly foreshadowed in it. A second more specific purpose is to relate the book to current thinking in the field of psychiatry. While the climate of epistemological thought has been changing and evolving throughout the world, the thinking of the author has undergone changes, precipitated especially by contact with some of the problems of psychiatry. I have had the task of teaching cultural anthropology to psychiatric residents, and have had to face such problems as are raised by the comparison between the variety of cultures and those hazily defined "clinical entities," the mental diseases which have their roots in traumatic experience.

This narrower purpose, to make the book relevant to psychiatry, is easier to achieve than the wider one of placing it in the current epistemological scene. And therefore I will attempt the psychiatric problem first, with this reminder to the reader—that the problems of psychiatry are after all shot through with epistemological difficulties.

Naven was written almost without benefit of Freud. One or two reviewers even complained about this, but I think that the circumstance was fortunate. My psychiatric taste and judgment were at that time defective, and probably a greater contact with the Freudian ideas would only have led me to misuse and misunderstand them. I would have indulged in an orgy of interpreting symbols, and this would have distracted me from the more important problems of interpersonal and intergroup process. As it was, I did not even notice that the crocodile jaw which is the gate to the initiatory enclosure is called in Iatmul *tshuwi iamba*—literally, "clitoris gate." This piece of data would really only confirm what is already implied when the male initiators are identified as "mothers" of the novices, but still the temptation to analyze the symbolism could have interrupted the analysis of the relationship.

But the fascination of symbol analysis is not the only pitfall in psychiatric theory. Perhaps even more serious are the distractions of psychological typology. One of the great errors in anthropology has been the naïve attempt to use psychiatric ideas and labels to explain cultural difference, and certainly the weakest part of the book is that chapter in which I tried to describe ethological contrast in terms of Kretschmer's typology.

No doubt more modern approaches to the problem of typology, such as Sheldon's work on somatotypes, are a great improvement upon the crudely dualistic system of Kretschmer. But this is not the point which concerns me. If Sheldon's typology had been available to me in 1935, I would have used it in preference to that of Kretschmer, but I would still have been wrong. As I see it today, these typologies, whether in cultural anthropology or in psychiatry, are at best heuristic fallacies, *culs de sac*, whose only usefulness is to demonstrate the need for a fresh start. Fortunately, I separated my dalliance with psychiatric typology into a single chapter; if this were not so, I would hardly permit the republication today.

But still the status of typology is undefined and crucial. Psychiatrists still hanker after classifications of mental disease, biologists still hanker for genera and species; and physiologists still hanker after a classification of human individuals which shall show a coincidence between classes defined by behavioral criteria and classes defined by anatomy. Lastly, be it confessed, I myself hanker for a classification, a typology, of the processes of interaction as it occurs either between persons or between groups.

This is a region in which problems of epistemology become crucial for the whole biological field, including within that field both the Iatmul culture and psychiatric diagnosis. There is an area of comparable uncertainty in the whole theory of evolution: Do species have real existence or are they only a device of description? How are we to resolve the old controversies between continuity and discontinuity? Or how shall we reconcile the contrast which recurs again and again in nature between continuity of change and discontinuity of the classes which result from change?

It seems to me, today, that there is a partial answer to these problems in the processes of schismogenesis which are analyzed in this book, but this partial answer could hardly have beeen extracted from that analysis when the book was written. These further steps had to wait upon other advances, such as the expansion of learning theory, the development of cybernetics, the application of Russell's Theory of Logical Types to communications theory, and Ashby's formal analysis of those orders of event which must lead to parametric change in previously steady-state systems.

A discussion of the relationship between schismogenesis and these more modern developments of theory is therefore a first step toward a new synthesis. In this discussion I shall assume that formal analogies exist between the problems of change in all fields of biological science.

The process of schismogenesis, as described in the book, is an example of progressive or *directional* change. And a first problem in evolution is that of direction. The conventional stochastic view of mutation assumes that change will be random, and that direction is only imposed upon evolutionary change by some phenomenon like natural selection. Whether such a description is sufficient to explain the phenomenon of orthogenesis—the long process of continuous directional change shown by the fossil record in ammonites, sea urchins, horses, asses, titanotheres, etc.—is very doubtful. An alternative or supplementary explanation is probably necessary. Of these the most obvious is climatic or other progressive change in the environment, and this type of explanation may be appropriate for some of the orthogenic sequences. More interesting is the possibility that the progressive environmental change might occur in the *biological* environment of the species concerned, and this raises questions of a new order. Marine organisms like ammonites or sea urchins can hardly be supposed to have much effect upon the weather, but a change in the ammonites might affect their biological environment. After all, the most important elements in the environment of an individual organism are (a) other individuals of the same species and (b) plants and animals of other species with which the given individual is in intense interactive relationship. The survival value of a given characteristic is likely to depend in part upon the degree to which this characteristic is shared by other members of the species; and, vis-à-vis other species, there must exist relationships—e.g., between predator and prey—which are comparable to those evolving interactive systems of attack and defense so grievously familiar in armaments races at the international level.

These are systems which begin to be closely comparable to the phenomena of schismogenesis with which this book is concerned. In the theory of schismogenesis, however (and in armaments races), an additional factor to account for the directedness of change is assumed. The direction toward more intense rivalry in the case of symmetrical schismogenesis, or toward increasing differentiation of role in complementary schismogenesis, is assumed to depend upon phenomena of learning. This aspect of the matter is not discussed in the book, but the whole theory rests upon certain ideas about processes of character formation—ideas which are also latent in most psychiatric theory. These ideas may be briefly summarized.

The order of learning to which I refer is that which Harlow has called "set-learning," and which I myself have called "deutero-learning." I assume that in any learning experiment—e.g., of the Pavlovian or the Instrumental Reward type—there occurs not only that learning in which the experimenter is usually interested, namely, the increased frequency of the conditioned response in the experimental context, but also a more abstract or higher order of learning, in which the experimental subject improves his ability to deal with contexts of a given type. The subject comes to act more and more as if contexts of this type were expectable in his universe. For example, the deutero-learning of the animal subjected to a sequence of Pavlovian experiences will presumably be a process of character formation whereby he comes to live as if in a universe where premonitory signs of later reinforcements (or unconditioned stimuli) can be detected but nothing can be done to precipitate or prevent the occurrence of the reinforcements. In a word, the animal will acquire a species of "fatalism." In contrast, we may expect the subject of repeated Instrumental Reward experiments to deutero-learn a character structure that will enable him to live as if in a universe in which he can control the occurrence of reinforcements.

Now, all those psychiatric theories which invoke the past experience of the individual as an explanatory device depend necessarily upon some such theory of high-order learning, or learning to learn. When the patient tells the therapist that, in her childhood, she learned to operate a typewriter, this is of no particular interest to him unless he happens to be a vocational counselor as well as therapist. But when she starts to tell him about the context in which she learned this skill, how her aunt taught her and rewarded her or punished her or withheld reward and punishment, then the psychiatrist begins to be interested; because what the patient learned from formal characteristics or *patterns* of the contexts of learning is the clue to her present habits, her "character," her manner of interpreting and participating in the interaction between herself and others.

This type of theory which underlies so much of psychiatry is also fundamental to the idea of schismogenesis. It is assumed that the individual in a symmetrical relationship with another will tend, perhaps unconsciously, to form the habit of acting as if he expected symmetry in further encounters with that other, and perhaps, even more widely, in future encounters with all other individuals.

The ground is thus laid for progressive change. As a given individual learns patterns of symmetrical behavior he not only comes to expect this type of behavior in others, but also acts in such a way that others will experience those contexts in which they in turn will learn symmetrical behavior. We have here a case in which change in the individual affects the environment of others in a way which will cause a similar change in them. This will act back upon the initial individual to produce further change in him in the same direction.

But this picture of schismogenesis cannot be true of Iatmul society as I observed it. Evidently, what has been achieved is only a one-sided picture of processes which, *if permitted*, would lead either in the direction of excessive rivalry between symmetrical pairs or groups of individuals or in the direction of excessive differentiation between complementary pairs. At a certain point, if these were the only processes involved, the society would explode. Of this difficulty I was already aware when I wrote the book, and I made an effort to account for the presumed dynamic equilibrium of the system by pointing out that the symmetrical and complementary processes are in some sense opposites of each other (page 193) so that the culture containing both of these processes might conceivably balance them one against the other. This, however, was at best an unsatisfactory explanation, since it assumed that two variables will, *by coincidence*, have equal and opposite values; but it is obviously improbable that the two processes will balance each other unless some functional relationship obtains between them. In the so-called dynamic equilibrium of chemical re-actions, the rate of change in one direction is a function of the concentration of the products of the inverse change, and reciprocally. But I was not able to see any such functional dependence between the two schismogenic processes and had to leave the matter there when the book was written.

The problem became totally changed with the growth of cybernetic theory. It was my privilege to be a member of the Macy Conference which met periodically in the years following the end of World War II. In our earlier meetings the word "cybernetics" was still uncoined, and the group was gathered to consider the implications for biology and other sciences of what we then called "feed-back." It was immediately evident that the whole problem of purpose and adaptation—the teleological problem in the widest sense—had to be reconsidered. The problems had been posed by the Greek philosophers, and the only solution they had been able to offer was what looked like a mystical idea: that the end of a process could be regarded as a "purpose," and that this end or purpose could be invoked as an explanation of the process which *preceded* it. And this notion, as is well known, was closely connected with the problem of the *real* (i.e., transcendent rather than immanent) nature of forms and patterns. The formal study of feed-back systems immediately changed all this. Now, we had mechanical models of causal circuits which would (if the parameters of the system were appropriate) seek equilibria or steady states. But *Naven* had been written with a rigorous taboo on teleological explanation: the end could never be invoked as an explanation of the process.

The idea of negative feedback was not new; it had been used by Clark Maxwell in his analysis of the steam engine with a governor, and by biologists such as Claude Bernard and Cannon in the explanation of physiological homeostasis. But the power of the idea was unrecognized. What happened at the Macy meetings was an exploration of the enormous scope of these ideas in the explanation of biological and social phenomena.

The ideas themselves are extremely simple. All that is required is that we ask not about the characteristics of lineal chains of cause and effect but about the characteristics of systems in which the chains of cause and effect are circular or more complex than circular. If, for example, we consider a circular system containing elements *A, B, C,* and *D*—so related that an activity of *A* affects an activity of *B, B* affects *C, C* affects *D,* and *D* has an effect back upon *A*—we find that such a system has properties totally different from anything which can occur in lineal chains.

Such circular causal systems must in the nature of the case either seek a steady state or undergo progressive exponential change; and this change will be limited either by the energy resources of the system, or by some external restraint, or by a breakdown of the system as such.

The steam engine with the governor illustrates the type of circuit which may seek a steady state. Here the circuit is so constructed that the faster the piston moves the faster the governor spins; and the faster the governor spins the wider the divergence of its weighted arms; and the wider the divergences of these arms the *less* the power supply. But this in turn affects the activity of the piston. The self-corrective characteristic of the circuit as a whole depends upon there being within the circuit at least one link such that the more there is of something, the less there will be of something else. In such cases the system may be self-corrective, either seeking a steady rate of operation or oscillating about such a steady rate.

In contrast, a steam engine with a governor so constructed that a wider divergence of the arms of the governor will *in-*

crease the supply of steam to the cylinder affords an instance of what the engineers would call "runaway." The feedback is "positive," and the system will operate faster and faster, increasing its speed exponentially to the limit of the available supply of steam or to the point at which the flywheel or some other part must break.

For the present purposes it is not necessary to go into the mathematics of such systems except to notice that the characteristics of any such system will depend upon timing. Will the corrective event or message reach the point at which it is effective at an appropriate moment and will the effect be sufficient? Or will the corrective action be excessive? Or too little? Or too late?

Substituting the notion of self-correction for the idea of purpose or adaptation defined a new approach to the problems of the Iatmul culture. Schismogenesis appeared to promote progressive change, and the problem was why this progressive change did not lead to a destruction of the culture as such. With self-corrective causal circuits as a conceptual model, it was now natural to ask whether there might exist, in this culture, functional connections such that appropriate controlling factors would be brought into play by increased schismogenic tension. It was not good enough to say that symmetrical schismogenesis happened by coincidence to balance the complementary. It was now necessary to ask, is there any communicational pathway such that an increase in symmetrical schismogenesis will bring about an increase in the corrective complementary phenomena? Could the system be circular and self-corrective?

The answer was immediately evident (page 8). The *naven* ceremonial, which is an exaggerated caricature of a complementary sexual relationship between *wau* and *laua*, is in fact set off by overweening symmetrical behavior. When *laua* boasts in the presence of *wau*, the latter has recourse to *naven* behavior. Perhaps in the initial description of the contexts for *naven* it would have been better to describe this as the primary context, and to see *laua's* achievements in headhunting, fishing, etc., as particular examples of an achieved ambition or vertical mobility in *laua* which place him in some sort of symmetrical

relationship with *wau*. But the Iatmul do not think of the matter this way. If you ask a Iatmul about the contexts for *naven*,
he will first enumerate *laua's* achievements and only as an
afterthought mention the less formal (but perhaps more profoundly significant) contexts in which *wau* uses *naven* to control that breach of good manners of which *laua* is guilty when
he presumes to be in a symmetrical relationship with *wau*.
Indeed, it was only on a later visit to Iatmul that I discovered
that when *laua* is a baby and is being held in *wau's* lap, if the
baby urinates, the *wau* will threaten *naven*.

It is also interesting that this link between symmetrical and
complementary behavior is doubly inverted. The *laua* makes
the symmetrical gesture and *wau* responds not by overbearing
complementary dominance but by the reverse of this—exaggerated submission. Or should we say the reverse of this reverse? *Wau's* behavior is a caricature of submission?

The sociological functions of this self-corrective circuit
cannot be so easily demonstrated. The questions at issue are
whether excessive symmetrical rivalry between clans will in
fact increase the frequency with which *lauas* act symmetrically
vis-à-vis their *waus*, and whether the resulting increase in frequency of *naven* will tend to stabilize the society. This could
only be demonstrated by statistical study and appropriate
measurement, which would be extremely difficult. There is,
however, a good case for expecting such effects inasmuch as
the *wau* is usually of a different clan from *laua*. In any instance
of intense symmetrical rivalry between two clans we may expect an increased probability of symmetrical insult between
members, and when the members of the pair happen to be
related as *laua* and *wau*, we must expect a triggering of the
complementary rituals which will function toward mending
the threatened split in society.

But if there exists a functional relationship such that excess
of symmetrical rivalry will trigger complementary rituals, then
we should expect to find also the converse phenomenon. Indeed, it is not clear that the society could maintain its steady
state without an excess of complementary schismogenesis
touching off some degree of symmetrical rivalry.

This too can be demonstrated with ethnographic data:

(1) In the village of Tambunum, when two little boys exhibit what looks to their age mates like homosexual behavior, the others put sticks in their hands and make the two stand up against each other and "fight." Indeed, any suggestion of passive male homosexuality is exceedingly insulting in Iatmul culture and leads to symmetrical brawling.

(2) As discussed in the book, while the *wau's* transvesticism is a caricature of the female role, the transvesticism of father's sister and elder brother's wife is a proud exhibition of masculinity. It looks as though these women are stating a symmetrical rivalry vis-à-vis the men, compensating for their normally complementary role. It is perhaps significant that they do this at a time when a man, the *wau*, is stating his complementarity vis-à-vis *laua*.

(3) The extreme complementarity of relationship between initiators and novices is always counterbalanced by extreme rivalry between initiatory groups. Here again complementary behavior in some way sets the scene for symmetrical rivalry.

Again we may ask the sociological question, whether these changes from complementarity to symmetry can be regarded as effective in the prevention of social disintegration; and, again, to investigate this question with the available examples is difficult. There is, however, another aspect of the matter which justifies us in believing that this oscillation between the symmetrical and the complementary is likely to be of deep importance to the social structure. What has been demonstrated from the data is that Iatmul individuals recurrently experience and participate in such shifts. From this we may reasonably expect that these individuals *learn*, besides the symmetrical and the complementary patterns, to expect and exhibit certain sequential relations between the symmetrical and the complementary. Not only must we think of a social network changing from moment to moment and impinging upon the individuals, so that processes tending toward disintegration will be corrected by activation of other processes tending in an opposite direction, but also we have to remember that the component individuals of that network are themselves being trained to introduce this type of corrective change in their dealings with each other. In one case we are equating

the individuals with the *A, B, C,* and *D* of a cybernetic diagram; and in the other, noting that *A, B, C,* etc., are themselves
so structured that the input–output characteristics of each will
show appropriate self-corrective characteristics.

It is this fact—that the patterns of society as a major entity
can by learning be introjected or conceptualized by the partici-
pant individuals—that makes anthropology and indeed the
whole of behavioral science peculiarly difficult. The scientist
is not the only human being in the picture. His subjects also
are capable of all sorts of learning and conceptualization and
even, like the scientist, they are capable of errors of conceptualization. This circumstance, however, leads us into a further set
of questions posed by communications theory, namely, those
questions which concern the *orders* of event which will trigger
corrective action, and the order of that action (considered as a
message) when it occurs.

Here I use the word "order" in a technical sense closely re-
sembling the sense in which the word "type" is used in Russell's Theory of Logical Types. This may be illustrated by the
following example. A house with a thermostatically controlled
heating system is a simple self-corrective circuit of the sort
discussed above. A thermometer appropriately placed in the
house is linked into the system to control a switch in such a
way that when the temperature goes above a certain critical
level the furnace is switched off. Similarly, when the temperature falls below a certain level the furnace is switched on. But
the system is also governed by another circumstance, namely,
the setting of the critical temperatures. By changing the position of a dial the owner of the house can alter the characteristics of the *system as a whole* by changing the critical temperatures at which the furnace will be turned on and shut off.
Following Ashby, I will reserve the word "variables" for those
measurable circumstances which change from moment to moment as the house oscillates around some steady temperature,
and shall reserve the word "parameters" for those characteristics of the system which are changed for example when the
householder intervenes and changes the setting of the thermostat. I shall speak of the latter change as of higher order than
changes in the variables.

The word "order" is in fact here used in a sense comparable to that in which it was used earlier in this essay to define orders of learning. We deal as before with meta relationships between messages. Any two orders of learning are related so that the learning of one order is a learning *about* the other, and similarly in the case of the house thermostat the message which the householder puts into the system by changing the setting is *about* how the system shall respond to messages of lower order emanating from the thermometer. We are here at a point where both learning theory and the theory of cybernetic systems come within the realm of Russell's Theory of Types.

Russell's central notion is the truism that a class cannot be a member of itself. The class of elephants has not got a trunk and is not an elephant. This truism must evidently apply with equal force when the members of the class are not things but names or signals. The class of commands is not itself a command and cannot tell you what to do.

Corresponding to this hierarchy of names, classes, and classes of classes, there is also a hierarchy of propositions and messages, and within this latter hierarchy the Russellian discontinuity between types must also obtain. We speak of messages, of meta-messages, and meta-meta-messages; and what I have called deutero-learning I might appropriately have called meta-learning.

But the matter becomes more difficult because, for example, while the class of commands is not itself a command, it is possible and even usual to give commands in a meta-language. If "Shut the door" is a command, then "Listen to my orders" is a meta-command. The military phrase, "That is an order," is an attempt to enforce a given command by appeal to a premise of higher logical type.

Russell's rule would indicate that as we should not classify the class of elephants among its members, so also we should not classify "Listen to my orders" among such commands as "Shut the door." But, being human, we shall continue to do so and shall inevitably be liable to certain sorts of confusion, as Russell predicts.

Returning to the theme which I am trying to elucidate—the general problem of the continuity of process and the discontinuity of the products of process—I will now consider how we might classify the answers to this general problem. Of necessity, the answers will be in the most general terms, but still it is of some value to present an ordering of thoughts about change as it must, *a priori*, occur in all systems or entities which either learn or evolve.[1]

First, it is necessary again to stress the distinction between change in the variables (which is, by definition, within the terms of the given system) and change in the parameters, i.e., change in the very terms which define the system—remembering always that it is the observer who does the defining. It is the observer who creates messages (i.e., science) about the system which he is studying, and it is these messages that are of necessity in some language or other and must therefore have *order*: they must be of some or other Logical Type or of some combination of Types.

The scientist's task is only to be a good scientist, that is, to create his description of the system out of messages of such logical typology (or so interrelated in their typology) as may be appropriate to the particular system. Whether Russell's Types "exist" in the systems which the scientist studies is a philosophic question beyond the scientist's scope—perhaps even an unreal question. For the scientist, it suffices to note that logical typing is an inevitable ingredient in the relationship between any describer and any system to be described.

What I am proposing is that the scientist should accept and *use* this phenomenon, which is, in any case, inevitable. His science—the aggregate of his messages about the system which he is describing—will be so constructed that it could be mapped in some more or less complex diagram of logical types. As I imagine it, each message of the description would have its

---

[1] This is not the place to discuss the controversies which have raged over the relation between learning and evolutionary process. Suffice it that two contrasting schools of thought are in agreement on a fundamental analogy between the two genera of process. On the one hand, there are those who, following Samuel Butler, argue that evolutionary change is a sort of learning; on the other hand, there are those who argue that learning is a sort of evolutionary change. Notable among the latter are Ashby and Mosteller, whose models of learning involve stochastic concepts closely comparable to the concepts of natural selection and random mutation.

location on this map, and the topological relationship between various locations would represent the typological relationship between messages. It is of the nature of all communication, as we know it, that some such mapping be possible.

But in describing a given system, the scientist makes many choices. He chooses his words, and he decides which parts of the system he will describe first; he even decides into what parts he will divide the system in order to describe it. These decisions will affect the description as a whole in the sense that they will affect the map upon which the typological relations between the elementary messages of the description are represented. Two equally sufficient descriptions of the same system could conceivably be represented by utterly different mappings. In such a case, is there any criterion by which it would be possible for the scientist to choose one description and discard the other?

Evidently an answer to this question would become available if scientists would use, as well as accept, the phenomena of logical typing. They are already scrupulous about the precise coding of their messages and insist upon singularity of referent for every symbol used. Ambiguity at this simpler level is abhorred and is avoided by rigorous rules for the translation of observation into description. But this rigor of coding could also be useful on a more abstract level. The typological relations between the messages of a description could also be used, subject to rules of coding, to represent relations within the system to be described.

After all, any modification of the signal or change in relationship between modifications of the signal can be made to carry a message; and by the same token any change in relationship between messages can itself carry a message. There is then no inherent reason why the various species of meta relationship between the messages of our description should not be used as symbols whose referents would be relationships within the system to be described.

Indeed, something like this technique of description is already followed in certain fields, notably in the equations of motion. Equations of first order (in x) denote uniform velocity; equations of second order (in x^2) imply acceleration; equations of third order (in x^3) imply a change in acceleration; and so on. There is moreover an analogy between this hierarchy of equations and the hierarchy of Logical Types: a statement of acceleration is *meta* to a statement of velocity. The familiar Rule of Dimensions is to physical quantities what the Theory of Logical Types is to classes and propositions.

I am suggesting that some technique of this sort might be used in describing change in those systems which either learn or evolve, and further that if such a technique were adopted, there would then be a natural basis upon which to classify answers to the problems of change in these systems: the answers would fall into classes according to the typology of the messages which they contain. And this classification of answers should coincide both with a classification of systems according to their typological complexity and with a classification of changes according to their *orders*.

In illustration of this, it is now possible to go back over the whole body of description and argument contained in this book and to dissect it on a generalized typological scale or map.

The book starts with two descriptions of Iatmul culture, in each of which relatively concrete observations of behavior are used to validate generalizations. The "structural" description leads to and validates eidological generalization, and a corpus of ethological generalization is validated by observations of expression of affect.

In the 1936 Epilogue, it is demonstrated that ethos and eidos are only alternative ways of arranging data or alternative "aspects" of the data. This, I believe, is another way of saying that these generalizations are of the same order or Russellian type. I needed, for obscure reasons, to use two sorts of description, but the presence of these two descriptions does not denote that the system described actually has a complexity of this dual nature.

One significant duality has however already been mentioned in this brief survey, namely, the duality between observations of behavior and generalization, and this duality, I believe, here reflects a special complexity in the system: the dual fact of learning and learning to learn. A step in Russellian typology inherent in the system is represented by a corresponding step in the description.

A second typological contrast in the description, which I believe represents a real contrast in the system described, is that between ethos–eidos on the one hand and sociology on the other. Here, however, the matter is less clear. Insofar as the total society is represented in native thought and communication, clearly this representation is of higher type than the representations of persons, actions, and so on. It would follow that a segment of the description should be devoted to this entity, and that the delimiting of this segment from the rest of the description would represent a real typological contrast within the system described. But, as the matter is presented in the book, the distinctions are not perfectly clear, and the idea of sociology as a science dealing with the adaptation and survival of societies is mixed up with the concept "society" as a *Gestalt* in native thought and communication

It is appropriate next to ask about the concept "schismogenesis." Does the isolating and naming of this phenomenon represent an extra order of complexity in the system?

Here the answer is clearly affirmative. The concept "schismogenesis" is an implicit recognition that the system contains an extra order of complexity due to the combination of learning with the interaction of persons. The schismogenic unit is a two-person subsystem. This subsystem contains the potentialities of a cybernetic circuit which might go into progressive change; it cannot, therefore, be conceptually ignored and must be described in a language of higher type than any language used to describe individual behavior—the latter category of phenomena being only the events in one or another arc of the schismogenic subsystem.

It is necessary, next, to note that the original description contained a major error in its typological mapping. The description is presented as "synchronic,"[1] which is more modern terminology may be translated as "excluding irreversible

---

[1] There is also a second sense in which anthropologists use the word "synchronic": to describe a study of culture which ignores progressive change by considering only a brief (or infinitesimal) span of time. In this usage, a *synchronic* description differs from a *diachronic* almost as differential calculus differs from integral calculus.

change." The basic assumption of the description was that the system described was in a steady state, such that all changes within it could be regarded as changes in variables and not in the parameters. In self-justification, I may claim that I stated that there must exist factors which would control the runaways of schismogenesis—but still I overlooked what is crucial from the present point of view: that the system must contain still larger *circuits* which would operate correctively upon the schismogeneses. In omitting to make this deduction, I falsified the whole logical typology of the description by not depicting its highest level. This error is corrected in the earlier part of the present Epilogue.

It is thus possible, at least crudely, to examine the scientific description of a system and to relate the logical typology of the description to the circuit structure of the system described. The next step is to consider descriptions of change preparatory to asking how a classification of such descriptions may be related to problems of phenomenal discontinuity.

From what has already been said, it is clear that we must expect statements about change to be always in a language one degree more abstract than the language which would suffice to describe the steady state. As statements about acceleration must always be of higher logical type than statements about velocity, so also statements about cultural change must be of higher type than synchronic statements about culture. This rule will apply throughout the field of learning and evolution. The language for the description of character change must always be of higher type than the description of character; the language to describe psychiatric etiology or psychotherapy, both of which impute change, must always be more abstract than the language of diagnosis. And so on.

But this is only another way of saying that the language appropriate for describing change in a given system is that language which would also be appropriate for describing the top typological level in a steady state system having one more degree of complexity in its circuits. If the original description of Iatmul culture, as put forward in the body of the book, had been a sufficient and correct description of a steady state, then the language of the additional statements about the larger circuits would have been precisely that language appropriate to describe *change* or disturbance in that steady state.

When the scientist is at a loss to find an appropriate language for the description of change in some system which he is studying, he will do well to imagine a system one degree more complex and to borrow from the more complex system a language appropriate for his description of change in the simpler.

Finally it becomes possible to attempt a crude listing of types of change and to relate the items of this listing to the general problem from which I started—that of the contrast between continuity of process and discontinuity of the products of process.

Take for a starting point a system S of which we have a description with given complexity *C*, and observe that the absolute value of *C* is for present purposes irrelevant. We are concerned with the problem of *change* and not at all with absolute values.

Consider now events and processes within S. These may be classified according to the orders of statement which must be made in the description of S in order to represent them. The crucial question which must be asked about any event or process within S is: Can this event or process be included within a description of S as a *steady state* having complexity *C*? If it can be so included, all is well and we are not dealing with any change that will alter the parameters of the system.

The more interesting case, however, is that in which events or processes are noted in S which cannot be comprised within the steady-state description of complexity *C*. We then face the necessity of adding some sort of meta description to be chosen according to the type of disturbance which we observe.

Three types of such disturbance may be listed: (a) Progressive change, like schismogenesis, which takes place in the values of relatively superficial and fast changing variables. This, if unchecked, must always disrupt the parameters of the system. (b) Progressive change which, as Ashby has pointed out, must always occur in the more stable variables (or parameters?) when certain of the more superficial variables are *controlled*. This must happen whenever limitation is imposed upon those superficial and fast-changing variables which previously were essential links in some self-corrective circuit. An acrobat must always lose his balance if he is unable to make changes in the angle between his body and his balancing pole.

In either of these cases, the scientist is driven to add to his description of S statements of higher order than those included within the previous description C.

(c) Lastly there is the case of "random" events occurring within the system S. These become especially interesting when a degree of randomness is introduced into the very signals upon which the system depends for its self-corrective characteristics. The stochastic theories of learning and the mutation-natural-selection theory of evolution both invoke phenomena of this kind as basis for description or explanation of change. The stochastic theories of learning assume random changes of some kind in the neurological net, while the mutation theory assumes random changes in the chromosomal aggregate of messages.

In terms of the present discussion neither of these theories is satisfactory because both leave undefined the logical typing of the word "random." We must expect, *a priori*, that the aggregate of messages which we call a *genotype* must be made up of individual messages of very various typology, carried either by individual genes or by constellations of genes. It is even likely that, on the whole, more generalized and higher-type messages are more frequently carried by constellations of genes, while more concrete messages are in general carried by individual genes. Of this, there is no certain knowledge, but still it seems improbable that small "random" disturbances can alter with equal frequency messages of whatever logical type. We must ask, then: what distribution of disturbance among the messages of various types do the proponents of these theories have in mind when they use the word "random"? These, however, are questions more specific than the broad terms of the present discussion and are introduced only to illustrate the problems which are posed by the new epistemology which is now evolving.

The problem of discontinuity now falls into place in the sense that it is possible to classify the principal types of process and explanation which crystallize around this phenomenon. Consider still the hypothetical system, S, and the description of this system whose complexity I am calling C. The first type of discontinuity is that relatively trivial case in which the state of the system at a given time is observed to contrast with its state at some other time, but where the differences are still such as can be subsumed within the terms of the given description. In such cases, the apparent discontinuity will be either an artefact resulting from the spacing of our observations in time, or will be due to the presence of on-off phenomena in the communicational mechanism of the system studied.

A less trivial case is proposed in considering two similar systems S_1 and S_2, both of them undergoing continuous changes in their variables such that the two systems appear to be diverging or becoming more and more different one from the other. Such a case becomes nontrivial when some extra factor is involved which may prevent later convergence. But any such factor will of course have to be represented in the description of the systems by messages of higher logical type.

The next category of discontinuity will include all those cases which involve parametric contrast. I have considered briefly above the types of ongoing process which must lead to parametric disruption and have noted that these are all instances in which the description of the system undergoing change must be of higher logical type than would have been the case in absence of such processes. This I believe to be true even in that vast majority of instances where the parametric disturbance leads to gross simplification of parameters after the disruptive change. Most commonly such disruptions will—in accordance with laws of probability—result in the "death" of the system. In a few cases some simplified version of S may persist, and, in still fewer cases, the parametric disruption will lead to the creation of a new system, typologically more complex than the original S.

It is this rare possibility that is perhaps most fascinating in the whole field of learning, genetics, and evolution. But, while in the most general terms it is possible to state with some rigor what sort of changes are here envisaged and to see the results of such progressive discontinuous change in, for example, the telencephalization of the mammalian brain, it is still totally impossible to make formal statements about the categories of parametric disturbance which will bring about these positive gains in complexity.

Here is the central difficulty which results from the phenomenon of logical typing. It is not, in the nature of the case, possible to predict from a description having complexity C what the system would look like if it had complexity $C + 1$.

This formal difficulty must in the end always limit the scientific understanding of change and must at the same time limit the possibilities of planned change—whether in the field of genetics, education, psychotherapy, or social planning.

Certain mysteries are for formal reasons impenetrable, and here is the vast darkness of the subject.

BIBLIOGRAPHY

W. R. Ashby. Design for a Brain. New York: Wiley, 1952.

———. Introduction to Cybernetics. New York: Wiley, 1956.

G. Bateson. "Bali: The Value System of a Steady State," in M. Fortes, ed., Social Structure. New York: Oxford University Press, 1949.

———. "The Message 'This Is Play,'" in Second Conference of Group Processes. New York: Josiah Macy Jr. Foundation, 1956. Pp. 145–242.

G. Bateson, D. D. Jackson, J. Haley, and J. H. Weakland. "Toward a Theory of Schizophrenia," Behavioral Science, I (1956), 251–64.

R. R. Bush and F. Mosteller. Stochastic Models for Learning. New York: Wiley, 1955.

S. Butler. Life and Habit. London: Fifield, 1878.

———. Luck or Cunning. London: Fifield, 1887.

H. von Foerster, ed. Cybernetics. 5 vols. New York: Josiah Macy Jr. Foundation, 1949–1953. (Transactions of the 6th, 7th, 8th, and 9th Conferences.)

C. P. Martin. Psychology, Evolution, and Sex. Springfield, Ill.: Charles C. Thomas, 1956.

L. F. Richardson. "Generalized Foreign Politics," *British Journal of Psychology Monograph Supplements*, No. 23 (1939).

J. Ruesch and G. Bateson. Communication: The Social Matrix of Psychiatry. New York: W. W. Norton, 1951.

B. Russell. Introduction to Wittgenstein, *Tractatus Logico-Philosophicus*. New York: Harcourt, 1956.

C. E. Shannon and W. Weaver. The Mathematical Theory of Communication. Urbana: University of Illinois Press, 1949.

C. H. Waddington. The Strategy of the Genes. London: Allen & Unwin, 1957.

H. Weyl. Philosophy of Mathematics and Natural Science. Princeton: Princeton University Press, 1949.

A. N. Whitehead and B. Russell. Principia Mathematica. Cambridge: Cambridge University Press, 1910.

N. Wiener. Cybernetics. New York: Wiley, 1948.

DIAGRAM OF KINSHIP TERMS USED IN THIS BOOK

A

B

KEY

A, Consanguineous terms (m.s.). B, Affinal terms (m.s.). ♂, male. ♀, female. =, marriage. |, descent. ⌐ siblingship.
Arabic numbers refer to individuals. Roman numbers refer to terms for patrilineal groups seen collectively.

1. nggwail.	7. iai.	10. mbora.	13. tshaishi.	16. tshuambo.	19. ianan.
2. iai.	8. ngai'.	11. wau.	14. nyarrun.	17. lando.	20. nian.
3. mbuambo.	9. ryame.	12. na.	15. tagwa.	18. nyanggai.	
4. naisagut.					21. laua.
5. tawonto.					22. kaishe-ragwa.
6. nondu.					

I. kaishe-nampa.
II. Own clan
III. lanoa-nampa or laua ryanggu.
IV. wau-nyame. (Son's iai nampa.)
V. iai-nampa. (Become towa-naisagut if EGO marries one of their women.)
VI. towa-naisagut. (Son's wau-nyame.)

INDEX GLOSSARY OF
TECHNICAL AND NATIVE WORDS

Adaptive function. When we say of some part of a functional system that it behaves in such and such a way in order to produce such and such a desirable effect in the system as a whole, we are attributing adaptive function; and we are verging upon teleological fallacy. But it is a cold plunge from this to realising that no cell or organ cares two hoots about our survival. We can avoid some of the dangers of teleology by acknowledging that undesirable effects occur not infrequently. We may also avoid treating the teleological theory too seriously by saying e.g. that the father plays no part in *naven* because there is no extra integration of the community to be got by further stressing of the links between father and son. Cf. pp. 27 and 211.

Affective and **cognitive.** These terms have been used loosely in the body of the book: affective to mean "pertaining to the emotions" and cognitive to mean "pertaining to thought". The terms are re-interpreted more critically on p. 275.

Affective function. The effect of some detail of cultural behaviour in satisfying or dissatisfying the emotional needs of the behaving individuals. Cf. p. 32.

agwi, islands of floating grass, which may attain considerable size and solidity. They are formed on lakes but float down to the Sepik when broken up. Crocodiles are said to live under them. Hence *wagan* (who are crocodiles) in their shamanic jargon refer to a house as *agwi.*

Alternation. Cf. p. 244 *et seq.*

angk-au, literally "potsherds", hence ancestral spirits symbolised by the old potsherds under the house. Cf. p. 45.

awan, the suffix of names given by mother's brother to sister's son. The literal meaning of this suffix is either "mask" or "old man". Cf. p. 43.

bandi, a novice or initiate; a young man; a member of initiatory grade *Ax* 3 or *Ay* 3. *-bandi* is a suffix on names given by *wau* to *laua* after initiation. Cf. p. 43.

b.s. (both speaking) means that a kinship term may be used by both sexes.

Calculating man. The individual in a community when seen from a purely economic point of view. This phrase need make no assumptions about the terms in which the man "calculates". Cf. p. 27.

Centripetal. The system of organisation of a group is said to be centripetal if it depends upon a single central authority or upon some form of hierarchy. Cf. p. 99.

Clan. I am using this word to mean a unilateral patrilineal group. Among the Iatmul there are between fifty and a hundred clans of which between ten and twenty are represented in any one village. In many cases the clans are paired and trace descent from a pair of brothers, one clan being called the "elder brother" of the other. Larger groups of clans also occur. See Moiety.

Cognitive. Cf. Affective.

Complementary. A relationship between two individuals (or between two groups) is said to be chiefly complementary if most of the behaviour of the one individual is culturally regarded as of one sort (e.g. assertive) while most of the behaviour of the other, when he replies, is culturally regarded as of a sort complementary to this (e.g. submissive). Cf. p. 176.

Configuration. Defined, p. 33.

Cyclothymia. A syndrome of characteristics of temperament described by Kretschmer (*Physique and Character*, 1925). This temperament is recognisable (*inter alia*) by a tendency to periodic variation between gaiety and sadness. It is believed to be associated with pyknic physique. Cf. p. 160.

Developmental psychology. The study of the changes which occur in the psychology of an individual as a result of growth or experience.

Diachronic. It is convenient to divide scientific anthropology into two major disciplines, the diachronic, which is concerned with the processes of cultural change, and the synchronic, which is concerned with the working of cultural systems at a given period. Synchronic anthropology ignores the historical origin of the details of culture. Cf. p. 3.

Diagonal dualism. A way of thought which groups persons or objects in pairs by regarding each member of the pair as the symmetrical counterpart of the other (p. 239). At the end of this book it is pointed out that "diagonal" is equivalent to "symmetrical" in the sense in which the latter term is applied to relationships. Cf. p. 271.

Direct dualism. A way of thought which groups persons or objects in pairs, regarding the members of each pair as mutually complementary (p. 239). For the equivalence between "direct" and "complementary" cf. p. 271.

Discrimination. The opposite of Identification (q.v.).

Dynamic equilibrium. A state of affairs in a functional system, such that, though no change is apparent, we are compelled to believe that small changes are continually occurring and counteracting each other. Cf. pp. 175 and 190.

Eidos. Cf. footnote on p. 25; and pp. 30 and 220.

Ethos. Cf. footnote on p. 2; and pp. 30, 32, 114, 119, 123 and 276.

External (or lateral) sanctions. I have used this term to mean those sanctions which are imposed upon a member of a group by persons outside that group. Cf. p. 98.

Formulation. I have used this word loosely as synonym for cultural premise.

Function. Cf. p. 26 *et seq.*

Grade. The Iatmul system of initiatory groups is not easily described in English and I have used the word "grade" somewhat unconventionally. For a description of the system see p. 245.

Homicide. A man who has killed another. Among the Iatmul, the successful killer of enemies is entitled to special ornaments, here called "homicidal ornaments".

iai, father's mother, father's mother's brother's daughter, all of the women of the same patrilineal clan as these. The same term (or *naisagut* or *tawonto*) may be applied to their brothers. For marriage between *iai* and *ianan*, cf. p. 88 *et seq.*

ianan, son's son (w.s.); father's sister's son's son (b.s.); the reciprocal of *iai* (q.v.).

iau, father's sister, real and classificatory.

Identification. Cf. explanation, p. 35.

Initiatory groups. Cf. diagram, p. 245.

Institution. Cf. criticism of this concept, p. 27.

Internal sanctions. I have used this term to mean only those sanctions imposed upon a member of a group by some other member (often an official) of the same group, and I have not used it in the psychological sense to refer to conscience, etc.

kaishe nampa, a collective term for the members of the patrilineal group into which ego's daughter is married.

kaishe-ndo, child's spouse's father; *kaishe-ragwa,* child's spouse's mother. These terms are also used between persons whose offspring might be expected to marry, e.g. sister's son's wife may be called *kaishe-ragwa.* The term *kaishe* is also used between partners who have exchanged ceremonial gifts of shell currency.

kamberail, an initiatory group, *By* 4 or *Bx* 4. Cf. p. 245.

kanggat, brother's child (w.s.); the reciprocal of *iau.*

kau, violence. Cf. p. 140.

kop, a general term for ancestral spirits which guard their descendants and help them in war. It includes both *mbwan* and *angk-au.*

lan, husband.

lando, sister's husband, sister's son's son, real or classificatory (m.s.). This term may also include men who might be expected to marry the speaker's sister, so that it is sometimes used instead of *ianan.*

lan men to! "husband thou indeed!". Cf. footnote, p. 82.

lanoa nampa, literally "husband people", a collective term for members of the patrilineal group into which the speaker's sister has married. Cf. explanation on p. 93.

Lateral sanctions. Cf. External sanctions.

laua, sister's child, sister's husband's father (m.s.), and other relatives classified with them. Cf. p. 94.

laua nyanggu, the patrilineal group which includes the speaker's classificatory *lauas.* Cf. p. 94.

laua-ianan, a synonym of *laua nyanggu.*

Leptorrhine. This term is loosely used to mean "having a narrow nose". Cf. p. 163.

Logic. Cf. footnote on p. 25, and p. 220.

mbapma, literally, a "line", especially a line of people side by side as distinct from a file. The word is also applied to a group defined in terms of alternating generations. In any patrilineal lineage there are two *mbapma,* one containing the members of ego's generation, his grandfather's generation and his son's son's generation; the other containing members of his father's generation and his son's generation.

mbora, mother's brother's wife, wives of all men classified as mother's brothers.

mbuambo, mother's father, mother's mother, mother's brother's son, mother's son's wife, father's elder brother. Cf. p. 39 for incorrect use of the term by a Iatmul.

mbuandi, a tree with brilliant orange-coloured fruit (*Ervatamia aurantiaca*).

mbwan, the spirit of an enemy who has been killed and buried under a standing stone or ceremonial mound. Such spirits are regarded as ancestors and are believed to help in warfare and in the increase of population.

mbwatnggowi, ceremonial dolls, which represent clan ancestral spirits associated with fertility. Cf. p. 233 and Pls. XXV and XXVII.

mbwole, a small ceremonial house for the boys, especially for initiatory grades *Ax* 5 and *Ay* 5.

mintshanggu, a mortuary ceremony. Cf. p. 47.

Moiety. Iatmul society is divided into two totemic groups or moieties called *nyowe* (Sun) and *nyame* (Mother). It is also divided into cross-cutting moieties for initiatory ceremonial (cf. p. 245). Membership in all these groups is determined by patrilineal descent.

Motivation. Cf. the examination of the terms "cognitive" and "affective" on p. 275.

m.s., man speaking; placed after the translation of a kinship term it means that that term is used in the given sense only by a man.

mwai, mythological beings analogous to *wagan* (slit gongs) and represented by masks with big noses. Cf. p. 233 and Pl. XXVIII b.

na, father's sister's child (m.s.); daughter's child (b.s.); reciprocal of *mbuambo.*

naisagut, wife's father, wife's mother, wife's brother's wife, wife's mother's brother. Cf. p. 91.

nambu wail, nambu=head, *wail*=crocodile. The term *wail* is used for any initiatory group (moiety, quadrant or generation group). The term "head of the crocodile" is applied to the senior member of a generation group in the initiatory system. Cf. p. 245.

nampa, people.

nasa, husband's sister's child; the reciprocal of *mbora.*

naven, a set of ceremonial customs of the Iatmul used to illustrate the theoretical analyses in this book.

ndjambwia, spikes set in the ground or magical devices designed to prevent trespass. Cf. p. 46.

ndo, man.

nemwan, big, great.

nggambut, a post set up beside the grave of a woman. Cf. p. 55.

nggambwa, vengeance; a fury who tears dead bodies to pieces. Cf. p. 57.

nggelakavwi, a tuberous epiphyte (*Myrmecodia* sp.). Cf. p. 59.

ngglambi. Cf. p. 54 *et seq.* and p. 140.

nggwail, father's father, father's father's sister, son's son, son's daughter. This term is also applied to the totemic ancestors of a clan.

nggwail-warangka. Cf. p. 39.

nggwat keranda. When a girl goes from her parents' house in marriage, her relatives decorate her. The ornaments constitute the *nggwat keranda,* a ceremonial present to the husband. Cf. p. 105.

nian, son, daughter, and (w.s.) father's sister's child. This is the ordinary word for children (pl. *nyanggu*) and is also used as a synonym for *nampa,* people.

nondu, daughter's husband, father's sister's husband, and husband's sister's husband; reciprocal of *naisagut.*

nyai', father, father's classificatory brother. Cf. *mbuambo* and *tshambwi nyai'.* This term is applied to father's sister in *naven.*

nyame, mother, mother's sister (real and classificatory); any woman of mother's clan; son's wife (m.s.). This term is applied to mother's brother in *naven.*

nyamun, elder sibling of the same sex, brother (w.s.); father's father. This term is also used between clans (q.v.) and initiatory grades (cf. p. 245), and is applied to elder brother's wife in *naven.*

nyanggai, sister, real or classificatory (m.s.); father's father's sister (m.s.); son's daughter (m.s.).

Glossary

Peripheral. The system of organisation of a group is said to be peripheral if it depends for its sanctions not upon a higher authority but upon the behaviour of other equivalent groups. Cf. p. 99.

Potlatch. Competitive ceremonial giving, characteristic of the Indians of the North-west Coast of British Columbia.

Pragmatic function. For criticism and subdivision of this concept, cf. p. 30 *et seq.*; also pp. 261 and 264.

Premise. Definition, p. 24.

pwivu. Cf. p. 46.

rite de passage, a ceremony accompanying or effecting a change of status.

Schismogenesis. Definition, p. 175.

Schizothymia. A syndrome of characteristics of temperament described by Kretschmer. (Cf. Cyclothymia.) Schizothymia is characterised (*inter alia*) by sudden and irregular changes from emotional anaesthesia to emotional hyperaesthesia.

Segmentation. A functional system may be said to be segmented when it consists of two or more parts, each of which is in some degree a repeat of the others, e.g. a system of clans or an earthworm.

Siblings. Persons of either sex, who have one or both parents in common.

Social psychology. Cf. tentative definition, p. 175.

Social structure. For the distinction between this concept and that of cultural structure, cf. p. 25.

Sociology. This word I have used in a specially restricted sense. Cf. pp. 30 and 34.

Standardisation. The process by which the individuals in a community are moulded to resemble each other in their behaviour. Cf. pp. 33, 34, 113, 255.

Structure. Cf. definitions of cultural and social structure, p. 25.

Symmetrical. A relationship between two individuals (or two groups) is said to be symmetrical if each responds to the other with the same kind of behaviour, e.g. if each meets the other with assertiveness.

Synchronic. Cf. Diachronic, and p. 256.

Syndrome. Used in this book as a collective term for the characteristics of a functional system. This term is used in medicine for the *additional* characteristics which a body acquires in a diseased condition.

tagail, a small ceremonial house of the boys, especially for members of *Bx* 4 and *By* 4. Cf. p. 136.

tagwa, woman, wife. This term is also applied jocularly to father's mother.

tambinyen, a partner in the opposite moiety, who is of the same generation group as the speaker, e.g. members of initiatory group *Ax* 3 are *tambinyanggu* of *Ay* 3.

tambointsha, feather tassels, a badge of homicide. Cf. pp. 46 and 72.

tavet, a rhythm on the slit gong used to summon a man or group.

tawontu, literally "wife-man", wife's brother, father's mother's brother, real or classificatory (m.s.). This term is the reciprocal of *lando* and is extended to include male *iai.*

timbut, a species of small lemon used in purificatory rituals, in magic, and as a purge. Cf. p. 155.

towa-naisagut, a collective term for the members of the wife's clan. Cf. p. 93.

Transvesticism. The wearing of women's clothes by men, or of men's clothes by women.

tshaishi, wife of elder brother, real or classificatory (m.s.).

tshambwi nyai', own father's younger brother.

tshat kundi, the act of addressing the sister's child as an ancestor. Cf. p. 44.

tshimangka, a species of fish; an act in initiation in which an initiator dances like a fish.

tshimbwora, a fringe of hanging strips of palm leaf used as a screen in initiation ceremonies. This word is used by *wagan* (shamans) for water. Cf. p. 47.

tshivla, long.

tshuambo, younger sibling of the same sex; son's son (m.s.); husband's younger brother.

Tshuosh, a tribe who live immediately north of the Iatmul.

tshugukepma, ceremonies in which the sister's children dance as the ancestors of their mother's brother's clan. Cf. p. 10.

tshumbuk, a pointing stick used in sorcery. This object is personified. Cf. pp. 60 and 72.

wagan. This term is used among the Central Iatmul in two senses: (*a*) for certain clan ancestral spirits who "possess" shamans and speak through their mouths; (*b*) for the sacred slit gongs used in initiation. Among the Eastern Iatmul the term is used only for the gongs and a separate term (*lemwail*) is used for the spirits. The Central Iatmul, however, regard the gongs and spirits as fundamentally the same. Cf. pp. 55, 233 and 236.

waingga, purchase. This term is also applied to the acquisition of a wife. Cf. p. 79.

warangka, father's father's father.

wau, mother's brother, son's wife's brother, and other relatives classified with them. Cf. p. 94.

wau-mbuambo, a collective term for the members of the clan of own or classificatory mother.

windjimbu, spirits which live in trees. Cf. p. 65.

w.s. Woman speaking; placed after the translation of a kinship term, it means that that term is used in that sense only by a woman.

yigen, beautiful, gentle, quiet. Cf. p. 141.

yigen kundi, literally "quiet singing". A mortuary ceremony performed by women. Cf. p. 156.

yivut. Cf. footnote, p. 129.

PLATES

The dancing ground of Palimbai in the flooded season. The *wompunau* or dancing ground is a long avenue down the centre of the village. In it stand the men's ceremonial houses, of which one is visible in the picture. The sides of the *wompunau* are lined with mounds on which coconut palms, crotons, bananas, etc. are planted to raise them above the flood level. Dwelling houses lie on both sides of the *wompunau,* hidden by this vegetation. The photograph was taken from the upper storey of another ceremonial house. In the immediate foreground are the tops of *Dracaena* plants growing on the *wak* or ceremonial mound belonging to this house. Behind these a woman is paddling a small canoe across the *wompunau.* In front and to the left of the far ceremonial house, its *wak* is visible.

PLATE I

A. *Naven* in Palimbai for a sister's son who had made a large new canoe. The two young men are *waus* (mother's brothers) and are dressed as old women, with tattered raincapes and bedraggled skirts. They hobble about the village supporting themselves by means of short paddles, such as are used by women in canoeing. (Men use paddles 10 feet to 12 feet long and stand while paddling.) The nearer of the two *waus* carries a white fowl in his hand to be presented to his *laua*.

B. *Naven* in Palimbai: the two *waus* searching for their *laua*

PLATE II

A. *Naven* in Palimbai. The *waus* are on the way to the canoe made by their *laua*. Suspended from the nose of the further *wau* is seen an "ornament" made of old sago.

B. *Naven* in Palimbai: one of the *waus*

PLATE III

A. *Naven* in Palimbai. One of the *waus* stumbling because of his pretended feebleness. The children crowd round laughing.

B. *Naven* in Palimbai. One of the *waus* has reached the canoe and falls into it with his legs widespread. The other is arriving miserably on all fours.

PLATE IV

A. A small *naven* in Kankanamun. The *laua nampa* (sister's children) had assisted in building operations. When the work was done, one of the *waus* dressed himself as a woman and is here seen dancing with a head taken from the enemy by one of his *lauas*. The head, a portrait modelled in clay on an enemy's skull, is an old one.

B. Initiation in Kankanamun. The novice is separated from his mother, and his *wau* ceremonially takes over her function and is called *nyame* ("mother"). The *wau* is seen carrying a novice pick-a-back as a real mother might carry her small child.

PLATE V

A. Women's *naven* in Kankanamun. A boy has made a new canoe, and the photograph shows his "mothers", father's sisters, and elder brothers' wives celebrating this achievement. This was only a small *naven* and the transvesticism is of a makeshift variety. The two "mothers" have discarded their skirts and *sit* in the canoe using women's paddles, i.e. they are "naked" and not transvestite. The fathers' sisters and elder brothers' wives, although they are using women's paddles, *stand* to paddle in the manner of men.

B. The same *naven*. The photograph shows the war-like gestures of the transvestite women. One of them is wearing a pubic apron made of leaves, a makeshift copy of that worn by men.

PLATE VI

A. Men's ceremonial house in Kankanamun. This was the finest ceremonial house in any of the river villages. The shingles of the sago-leaf thatch are alternately yellow and brown, giving a decorative pattern. In the gable front four little windows are visible, with an enemy's skull in each. Above these windows there is a grotesque face of which only the teeth and crescent-shaped nose ornaments are visible in the photograph. In front of the building is the *wak* or ceremonial mound on which dead enemy bodies and captives are laid. At the foot of the ladder is a little tree of the species called *wani*, a totemic ancestor of the clan which owns this end of the ceremonial house. The building is 130 feet in length.

B. A man debating. The man was posed for this photograph. He stands beside the elaborately carved debating stool, and holds in his hand the three bundles of leaves with which he beats the stool during his speech. Photograph taken in a ceremonial house in Malingai village.

PLATE VII

A. The corner of a ceremonial house in Palimbai. The photograph shows one of the posts which support the floor. The top of the post is carved to represent a *windjimbu* or wood spirit.

B. Raising the ridge pole for one of the ceremonial houses in Palimbai. A scaffolding has been constructed around the supporting post. The horizontal poles in this scaffolding are not tied down, so that they can be removed to allow the ridge pole to be raised. The horizontals are replaced under it at each step. The ridge pole is raised by men beside it on the scaffolding, and by others below who push on a cross-piece tied to a bamboo shaft. The shaft is attached to the ridge pole at the top by a loop of cane. Such a work is the duty of all the affinal relatives of the clan whose ceremonial house is being constructed. The work would be accompanied by beating of gongs and playing of flutes.

PLATE VIII

A. A pair of slit gongs in Mindimbit. The photograph shows only
the elaborately carved ends of the gongs. The carving on both these
specimens represents the heads of prawns with their rostra. The
interest in noses gives a special importance to such totemic ancestors
as possess conspicuous noses, e.g. sago weevils, prawns, etc. On the
nearer gong, at the base of the rostrum, a frog is represented with its
hind leg extended in a long wavy line.

B. Initiation in Kankanamun. The initiators are waiting in two lines armed with sticks.
The novice will enter through the leaf screen at the back of the picture and will run the
gauntlet of the initiators. Actually, the novice's father will come with him and will
protect the back of the novice, allowing blows to fall on his own back. This beating of
the father is analogous to a beating of the novice, since both are of the same initiatory
moiety.

PLATE IX

Initiation in Malingai. The novice is lying prone on an inverted canoe clasping his mother's brother who acts as a comforter and "mother". An initiator of the opposite moiety to the novice is cutting the latter's back with a small bamboo blade. In the foreground is a bowl of water with swabs of fibre to wipe away the blood. The white and black paint on the faces of the two men is a privilege of those who have killed a man, and is worn on all ceremonial occasions. The band of opossum fur worn by the initiator is also a badge of homicide.

PLATE X

A. Initiation in Kankanamun. The novice is sitting in the lap of his mother's brother while initiators cut crescents above the novice's nipples. The initiator with bird-of-paradise feathers in his hair is doing the cutting, while others observe his craftsmanship and hold the novice.

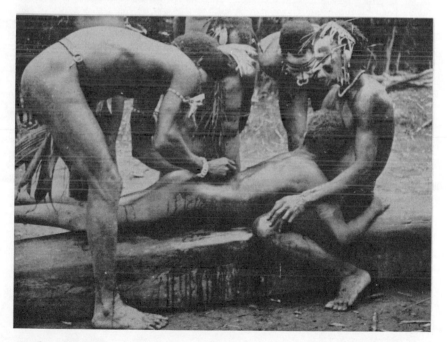

B. The same initiation: the cutting of the novice's back. Other initiators watch the operation.

PLATE XI

Initiation in Komindimbit village: bullying the novice. For about a week after his back has been cut, the novice is subjected every morning to a series of bullying ceremonials. He is made to squat like a woman while masked initiators maltreat him in various ways. The particular incident shown here is a recent innovation. The mask worn by the initiator was made by people in the mountains at the head of the Yuat River. This mask somehow—probably as loot—came into the possession of Lower Sepik natives. A party of returned Iatmul labourers stole the mask on their way to Komindimbit, and it was adopted into the local initiation ceremonies; it was referred to as *tumbuan* (the ordinary pidgin word for a mask). The *tumbuan* divines for theft. The cow's bone (acquired on some plantation) which is on the ground in front of the figure's knees is held up on a string in front of the novice's face. The *tumbuan* then says, addressing the novice in the feminine second person singular: "Are you a child that steals calico?" He then moves his arm causing the bone to swing, which indicates an affirmative answer to the divination; and he slaps the novice. The same question is asked about yams, bananas, tobacco, etc., etc. In each case an affirmative answer is obtained and the novice is slapped.

PLATE XII

A. Initiation in Komindimbit: an initiator expressing scorn for a novice by rubbing his buttocks on the latter's head. This gesture is presumably quite distinct from the gesture of the *wau*, who shames himself by rubbing his buttocks on his *luau*'s leg. (The dog in this picture belonged to the ethnographer and was not native.)

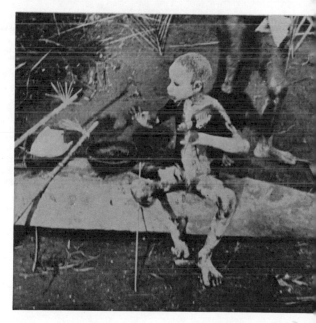

B. Initiation in Komindimbit: a novice eating a meal. He is made to wash before he eats, and even then must not touch his food with his hands, but must pick it up awkwardly and miserably with a pair of bamboo tongs or a folded leaf. This was the youngest of the novices initiated in this ceremony. The parts of his body which have been cut are wet with oil and the remainder of his body is smeared with clay.

PLATE XIII

Initiation in Komindimbit. Every day for nominally five days but actually for a rather longer period, the novices were bullied during the morning and allowed to rest in the afternoon. This photograph shows them resting, sitting or lying on mats on the ground in the ceremonial house. The initiators sit on raised platforms above the novices whom they are initiating. Down the centre of the ceremonial house the slit gongs are arranged in pairs, and initiators of the opposite moiety are sitting on the platforms on the far side of the house. Below them on the ground, hidden by the gongs, are the novices whom they are initiating. A pair of flutes is lying on the central gongs, and in front of the nearest house-post there is a pair of water drums. Feather headdresses, etc., belonging to the initiators are suspended on hooks from the beams.

PLATE XIV

A. Dwelling house in Kankanamun village

B. Interior of house in Kankanamun village. The foreground is filled by mosquito bags made of plaited rushes. Usually, a husband has his own bag and each wife has a bag. A separate larger bag, e.g. that on the extreme right of the photograph, is provided for the children, who sleep together. Suckling children sleep with their mothers. Suspended in the house are a number of baskets and patterned string bags imported from the Tshuosh people. The camera was set up on the floor in the clear space at the front end of the house, and there is a corresponding clear space beyond the mosquito bags at the far end.

PLATE XV

A. Women of Mindimbit going along the banks of the Sepik in small canoes to their fish traps

B. Women of Palimbai fishing in a lake. The circular net is held by two women who stand in the prows of two small canoes. Each canoe is propelled by a woman sitting in the stern. Such a net, called *djura*, is used to catch the fish which hide under small islands of floating grass. The net is pushed under the edge of the island and then raised. A small patch of grass is sometimes simply lifted up in the net and then thrown out.

PLATE XVI

A. Woman of Mindimbit village with child, the latter heavily ornamented with shells, as is usual even in everyday life. Both woman and child are rather shy of the ethnographer.

B. Woman of Kankanamun village with child. She did not know that she was being photographed.

PLATE XVII

A. An audience of women and children watching a spectacle staged by men. The young women and especially the children are heavily ornamented for the occasion. The spectacle in this case was a *wagan* dance.

B. Old men of age grades *By* 2 and *Bx* 2 dancing as *wagan*. *Wagan* are spiritual beings mythologically connected with water. Their voice is represented by the beating of slit gongs, a secret of late initiation. The men impersonating *wagan* are here seen prodigiously decorated, dancing with a *djura* (fish net, cf. Plate XVI B) in which the *wagan* were raised from water in their mythological origin. The net contains an imitation fish made of wood which is tied to the bottom of it.

PLATE XVIII

A. Woman painted and ornamented for *tshugu-kepma* procession, a public ceremonial in which the women march in a column up and down the dancing ground. They celebrate the ancestors of their *wau-nyame-nampa* (maternal clan). For such an occasion the women wear, in addition to their best skirts and other ornaments, a number of decorations which are usually only worn by men. In the photograph the headdress of casso-wary skin and the bobbing fans set with white fowls' feathers, are both of them normally worn by men, as also are the shell rings which the woman is wearing on her arm. The paint on her face is yellow ochre and not the white paint worn by homicides.

B. In front of the *tshugu-kepma* procession (cf. *supra*) a group of women perform a dance rep-resenting an incident in the mythology of the maternal clan. The figure in the left front of the picture is a woman wearing a faceless mask shaped to represent the Aibom mountain. Part of a second mask representing the Tchambuli mountain is visible on the extreme right. These two mountains are a conspicuous feature of the Sepik fens. All pieces of land are mythologically supposed to be floating islands, and here we see a woman steering the floating mountain with a long paddle such as is normally used by men. Her gesture, too, is reminiscent of male pride.

PLATE XIX

A. A great man of Palimbai had died the night before and was buried in the early morning. Later in the morning, the figure here shown was constructed to represent him. The head of the figure was an unripe coconut and the body was made of bundles of palm leaves. Spears were set up against it with their points stuck in the figure to mark the places where he had been wounded in fight, and other spears were stuck into the ground beside the figure for those which he had dodged. A series of vertical spears of which only two are visible in the photograph was set up in front of the figure according to his achievements. The figure itself was ornamented with shells, etc. Six sago baskets were suspended from its right shoulder to represent his six wives, a string bag was suspended on the left shoulder representing his skill in magic. A number of sprigs of ginger in its headdress represented persons whom he had invited to the village so that other people could kill them. In the right hand of the figure (not visible in the photograph) was a dry lump of sago—since it was said that in his lifetime he had once killed a bird by throwing a lump of sago at it. The branch of *timbut* (lemon) set in the ground beside the figure is symbolic of his knowledge of mythology.

B. The figure was set up by the initiatory moiety of which the dead man had been a member. Later, members of the other moiety came and took away the various symbols one by one. In this photograph a man is seen taking away one of the spears denoting a wound, claiming that he has had a similar wound. On the ground by the further foot of the figure is a broom and a pair of the boards used for picking up rubbish. These objects are symbolic of the work which the dead man had done in cleaning the ceremonial house during his lifetime.

PLATE XX

A. The row of spears in front of the mortuary figure shown in Plate XX. The figure is seen in the background. In front of it there are (*a*) eight spears for men he had killed, (*b*) ten spears for men killed by other members of the canoe on which he had served in the prow, (*c*) ten spears for wild pigs which he had killed: these are marked with pieces of banana leaf, and (*d*) (not visible in the photograph) spears for the crocodiles and flying foxes which he had killed.

B. Figures in *mintshanggu* mortuary ceremony, Palimbai. The skulls of the dead have been cleaned and portraits modelled upon them in clay. These are set up as the heads of the figures, which stand on a platform suspended from the roof of the house. Both the figures and platform are prodigiously ornamented. In front of the two figures is a portrait skull of the man whose mortuary figure is shown in Plate XX. The ceremony takes place at night, when the ancestral songs of the deceased are sung and men concealed under the platform play on the clan flutes.

PLATE XXI

Man of Mindimbit village of the proud, excitable, dramatising personality which is preferred in men. This man was indeed considered by the Iatmul to be somewhat too unstable. He was an enthusiastic, but confused and unreliable, informant. His posture shows the male reaction to the camera (cf. Plate XXVI). His hand grasps his lime stick, ready to scrape it on the mouth of the lime box, to produce a loud noise expressive of pride or anger.

PLATE XXII

A. Mali-kindjin of Kankanamun, whose personality is described in the chapter on "Preferred Types"

B. A man of Tambunum village

PLATE XXIII

A. Native of Mindimbit. A man of the discreet type and slightly pyknic physique: a very intelligent, humorous man and one of my most knowledgeable informants, to whom much esoteric knowledge had been entrusted because his elders had no fear of his declaiming their secrets in a sudden fit of temper. The pendants on his lime stick are a tally of the men he has killed.

B. Tshimbat, a native of Kankanamun, a product of culture contact

PLATE XXIV

The portrait skull of a woman. She was a native of Kankanamun who died some three generations ago. Her skull was cleaned, exhibited at mortuary ceremonies, and finally buried as is customary. But as she was considered to be strikingly beautiful, the men later dug up her portrait (and probably substituted another skull in the grave). Since then, her skull has been used in *mbwatnggowi* ceremonies (cf. Plate XXVII). Her long nose was especially admired. In the photograph, the breast is a half coconut shell.

PLATE XXV

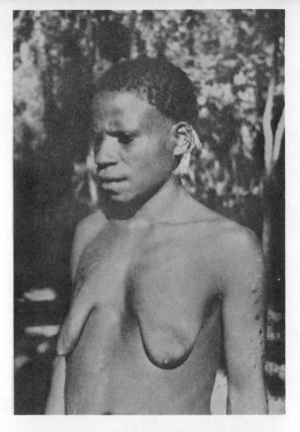

A and B. Two women of Kankanamun showing modesty before the camera (cf. Plate XXII)

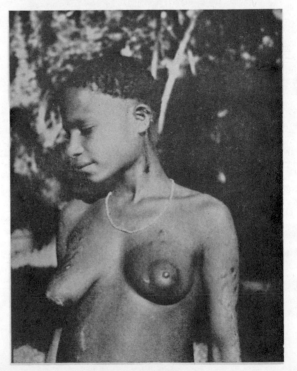

PLATE XXVI

A *mbwatnggowi* figure, Kankanamun. A portrait skull of a dead man, chosen for
his good looks, is set up as the head of a doll and prodigiously ornamented. The
white object above the head is a European plate which, being brighter and whiter,
is used instead of a Melo shell. The shell ornament is hung as an extension to the
nose. The ceremony is an initiatory secret, but is also said by a few men to promote
the prosperity and fertility of the village. The dance takes place at night behind a
screen of totemic banana leaves. The men lift up the dolls by the poles on which
they are constructed, raising them high above the screen so that the women, standing as an audience, see the figures as Jack-in-the-boxes. Each clan has its
mbwatnggowi, and the principal interest of the natives in the ceremony lies in its
spectacular nature and its esoteric origins rather than in its effect upon fertility.

PLATE XXVII

A. Figures of *wagan* with enormous noses, in Jentschan. This photograph shows the setting of the dancing ground for *wagan* dances. In the background to the right, the front of the ceremonial house is visible. In front of this is a screen of banana leaves through which the dancers impersonating *wagan* will come. The two figures in front of this screen have been hastily and secretly made ready and ornamented during the two days preceding the dance. They were set up in the dance ground in the small hours of the morning, so that the women woke to find the dancing ground prepared. Each figure consists of a post set at a high angle in the ground; to the top of this post a face is attached. The face is an enormous loop—the nose—near the base of which two half coconut shells are fixed to represent eyes. From the junction of the face and post a sago frond is suspended, and slopes forward on to the ground. The blades of the leaflets of this frond have been removed, and on to the mid-ribs of the leaflets large orange-coloured fruits have been threaded. This sago leaf is described as the *ioli* (raincape) of the figure. The post itself is entirely covered with totemic ancestral plants, one of the figures being decorated with ancestors of the Mother moiety, the other with ancestors of the Sun moiety. In front of the two figures, a rectangle is enclosed with white palm fronds. This is the "lake" in which the *wagan* will dance.

B. The head of a *mwai* mask, Kankanamun. *Mwai* are extinguisher-shaped masks, and the ceremony in which they perform is held in the dancing ground of the *tagail* (junior ceremonial house). The ceremony is a junior analogue of the *wagan* dances. The photograph shows the carved face detached from the extinguisher-shaped framework. The nose is extended into a long appendage ending in a snake's head. The face is made of wood, to which shells are attached by means of clay worked with lime and oil. The face is hanging in a dwelling house with various shell ornaments in the background.

PLATE XXVIII